DREAM

J. A. Hadfield was bo
parents. He was educ
ceeded to Oxford Univ
Arts, he read theolog ... At Oxford he fell under the influence of William McDougall, the psychologist, under whom he did some research work on the relation of anthropology to psychology. The deeper interest in abnormal psychology thus gained led him to take his degrees in Medicine at Edinburgh University with a view to the treatment of nervous disorders, then in its infancy. During the First World War he served as Surgeon-Lieut. R.N. until, towards the end of the war, he was transferred to the Army, where he worked once more with McDougall as Neurologist in the War Hospital at Oxford. Between the wars he practised in Harley Street in psychological medicine, was appointed Lecturer in Psychopathology and Mental Hygiene in London University, and was one of the founder members of the Tavistock Clinic, later becoming its first 'Director of Studies' for the training of doctors in psychological medicine. In the Second World War he was Officer in Charge of Division and Director of Studies at the 41st General (Neuropathic) Hospital. J. A. Hadfield died in 1967.

Among his numerous writings were *Psychology and Morals* (1920), *Psychology and Mental Health* (1950), *Mental Health and the Psychoneuroses* (1952) and *Childhood and Adolescence* (1962), a Pelican book.

Also available as a Pelican:

CHILDHOOD AND ADOLESCENCE

DREAMS AND NIGHTMARES

J. A. HADFIELD

PENGUIN BOOKS

Penguin Books Ltd, Harmondsworth, Middlesex, England
Penguin Books Inc., 7110 Ambassador Road, Baltimore, Maryland 21207, U.S.A.
Penguin Books Australia Ltd, Ringwood, Victoria, Australia

—

First published 1954
Reprinted 1954, 1961, 1962, 1964, 1966, 1967, 1969, 1971, 1973

—

—

Made and printed in Great Britain
by Cox & Wyman Ltd, London, Reading and Fakenham
Set in Monotype Scotch Roman

CONTENTS

Contents

EDITORIAL FOREWORD

THERE must be something a little unusual about people who have no interest in dreams. In themselves, dreams are often extremely good anecdotes. A lull occurs in the conversation. Someone says 'I had a funny dream last night', and all eyes and ears twist towards the speaker.

Dreams can be of intellectual interest, too. They offer a quick way in to some of the problems of philosophy, to the problems of psychical research, and perhaps the quickest way in of all to problems concerning the most curious regions of the human mind.

Few people have access to the mysteries of the psychoanalytic couch, but everyone, or almost everyone, has access to his dreams. And in our dreams we have at our disposal some of the most important facts which are critical for one of the great scientific issues of our age.

Those concerned with this curious region of the mind – psychologists and psychiatrists – are divided into two camps, with some psychologists and some psychiatrists in each. Members of the one camp are quite well satisfied that in its broadest outlines the psychoanalytic conception of the mind, as developed by Freud, his followers and successors, is substantially correct. In the opposite camp are the experts who range from those holding, at one extreme, that the conception is absolute nonsense to those who hold, at the other extreme, that the conception is, well, not quite scientific.

Both camps are quite properly represented in the series of Pelicans to which this volume belongs. Frieda Fordham has presented the case for analytic psychology as developed by Jung. Freud himself, Adler, and others, have also been represented. In this group belongs the present volume, though Dr Hadfield has the distinction of being able to present the psychoanalytical point of view without committing himself to the special doctrines of any particular school. He has the added distinction of offering in this volume an original contribution to the general theory – the biological theory of dreams.

The opposite camp is represented in this series by Dr Eysenck who, in his *Uses and Abuses of Psychology*, presents a powerful though not intemperate case against the psychoanalytic thesis. The dispute between the two camps is in some ways rather curious. One side says: Psychoanalytic doctrines are *true*. The other side says: Psychoanalysis is *unscientific*.

Clearly, these two statements are not inconsistent. In fact, the moderate critic of psychoanalytic theory takes the line: What you say may be true – in the sense in which the pronouncements of the poet, the prophet, or the seer may be true. Your interpretations of dreams and of psychoneurotic symptoms may display wisdom, insight, or even inspiration, but they have not the quality of established scientific fact.

All this might be accepted by the psychoanalyst who would not, nevertheless, wish to be dismissed with the poets and the mystics. His reply tends to take the following line: Psychoanalytic theory has not *yet* the status of established scientific fact. It has, however, the status of any really big scientific hypothesis.

It is agreed on all hands that any scientific theory begins as an hypothesis, a hunch, or in plain language a more or less shrewd *guess*. The function of scientific method is to test the guess by reference to observable fact. This takes time. Between the first conception of the hypothesis and its final refutation or validation the hypothesis has an ambiguous status. A general sense of the direction in which the evidence is pointing may sustain a not irrational confidence in the hypothesis even though the evidence cannot be set out in detailed statistics and tables. This is the present position of psychoanalytic theory. Freud may yet be proved to be the world's greatest guesser, and to have had all the best hunches. Before we know whether this be so or not, there must be years and years of tedious and pedestrian labour. That was the case with Newton and with Darwin.

However, it is not the function of the Editor of this series to intervene in the discussion. It is not even his function to be the Chairman of a Brains Trust, who dangles the succulent worm of compromise between the noses of the disputants. His function is that of the Announcer.

So here, readers, is Dr Hadfield, to tell you about your dreams and your nightmares.

C. A. MACE

PREFACE

MAN was ever an explorer. He loves to penetrate the trackless forest; to dive to the unseen depths of the sea; to scale mountains, and to delve into caves of the earth, in whose recesses he finds relics of earlier habitations, and in the ledges of which he discovers the bones of creatures which have long since disappeared from the surface of the earth. Yet in each one of us there are relics of bygone ages far beyond history and the human imagination, and in the depths of the unconscious mind realms of mystery far more fascinating than are to be found in the caves of the Mendips, traces of a remote ancestry, or archaic modes of thought, and of a world of imagination endowed with richer potentialities than are to be found in the treasures of the East.

All these are revealed to us night after night in our dreams. We go to the depths of the South American forests or to the plains of Australia to study the aborigines; but in the phantasmagoria of our dreams we are night after night confronted with visions more fantastic than anything to be seen in the strange rituals and rites of these primitive tribes. Night after night we have the excitement of big-game hunting of animals far more ferocious than those met with in Africa, and we are fortunate if we do not end by being the hunted! For these creatures of our dreams are ever breaking through the barriers of the unconscious mind to wake us up in terror. Or if wishing to escape from the monotony of the day we long for a haven of bliss, dreams provide us with visions of joy more ecstatic than those of the Arabian Nights, and allow us to partake of experiences such as we have never enjoyed nor are ever likely to enjoy – only to wake us to the realities of our daily life. Yet these glimpses of the ineffable give us a respite and permit us momentarily to find peace in an oasis more welcome than that of the desert, so that we can go forward more joyfully with our everyday task.

But as we pursue the study of these dreams, we suddenly discover that they not only reveal to us glimpses of our archaic ancestry, to which Jung has called our attention, and gratify the unfulfilled wishes of our lives, as Freud has taught us, but that hidden behind their fantastic visions they are full of shrewd common sense even in dealing with the practical affairs of everyday life; they contain flashes of insight which enable us to solve problems even more effectively than our lumbering reason, and reveal to us characteristics in ourselves which surprise and sometimes shock us. Indeed we discover that, unknown to us, dreams are busying themselves throughout the night with those problems of our lives which in the day we find too much for us, and by recapitulating the events of the day help towards a solution of those problems. Dreams rightly interpreted may thus act as a guide when we are lost in bewilderment in those trackless forests of the imagination, and help us to overcome difficulties even when we have abandoned them as insurmountable.

That is the theme of this book, and it is to that realm of the imagination that we would lead the reader. It is our aim to show, however inadequately, the sense that lies concealed in the apparent nonsense of the dreams, and perhaps to induce some readers to take a further interest in the fascinating, not to say amusing task of unravelling these mysteries of our nightly visions.

In the first chapter I attempt to give a brief account of the popular views, to be followed by those of the modern schools, especially of Freud, Jung, and others. My own views of the origin and nature of dreams, here described in the chapters on 'Dreams as Biological' and 'Nightmares', were originally given in a paper (previously unpublished) before the British Psychological Society more than twenty years ago. Not that I claim that these views are entirely original; indeed they are based on the pre-existing views, the popular views as well as the more scientific; but my theory, I think, gives to dreams a more important biological significance than previously held views.

If I have any claim to write on the subject, it is that while I have been dealing with the dreams of patients for the last thirty-six years and more, I do not belong to any of the established schools, and am perhaps therefore in a more favourable position to appreciate what is of value in each of them. My aim is to give the general reader the facts of the case as far as space will allow, with a few arguments for and against the various theories, and let him make up his own mind on the matter of dream interpretation. I recognize that for anyone to write a survey of this kind is, from one point of view, a thankless task, for in so far as he feels compelled to criticize some aspects of the established theories, he makes himself unpopular with all the schools which hold rigidly to their theories. But that is a risk we must take if we are to keep an open mind and strive to reach the truth.

If my description is inadequate, as indeed it will be in so short a summary, I can only plead that this book is merely an *introduction* to the study of the subject. I crave in advance, therefore, the indulgence of the reader for any such inadequacy, and the pardon of the adherents of the established schools for any unintentional misrepresentation of their views which is almost bound to occur in the attempt to give so brief a presentation of the subject.

For further enlightenment the reader must be referred to the works of the authorities themselves. Freud's book on *The Interpretation of Dreams* is a classic on which the views of other modern writers have been largely based, although they may in many respects disagree with him. Jung's views are not so specifically set out, but are found scattered throughout his writing. For a start, I should suggest Beatrice Hinkle's excellent Introduction to Jung's *Psychology of the Unconscious* (the book itself is too difficult for the beginner and hard going even for the expert), Jung's *Analytical Psychology*, and his *The Integration of the Personality*, and finally his *Modern Man in Search of a Soul*. There are, of course, other books on dreams and other theories, but to attempt to give a description of these would only confuse the reader, whereas the account we have given will be sufficient to establish the general principles of dream interpretation.

It is, perhaps, necessary to mention that all the dreams given as illustrations in this book are authentic dreams, mostly of my patients, unless otherwise stated, and my thanks are due to them for permission to use them. I have also to thank Dr E. A. Bennet, Dr Alice Buck; Mrs K. M. Goldney for help in particular chapters, Miss Beryl Eeman for reading through the whole in typescript and for many helpful suggestions; and most of all Professor Mace, the Editor of this series, for many helpful criticisms and suggestions which I have most gladly incorporated.

Harley Street, London
September 1952

J. A. HADFIELD

PART ONE

HISTORICAL SURVEY

CHAPTER 1

INTRODUCTION: POPULAR AND PHILOSOPHIC THEORIES

SEVERAL of us were seated at breakfast at a country house party one morning when one remarked that he had had a most peculiar dream the night before: 'I dreamt that I was in Trafalgar Square, and as I looked, all the statues, the lions, Nelson and all, suddenly came to life.' This got us talking about dreams. Asked what he dreamt about, one said: 'Oh! the usual dream – trying to catch a train and only just managing it.' Another said laughingly that he often had that same dream, but always just missed it! Another said she only dreamt of trifling things that had happened the day before, whilst another woman said: 'All I dreamt about was the word "Largs" – but why I should dream of that I do not know: it is a place in Scotland which I have not been to for twenty years, and then only for one night, crossing over to Ireland.' Our host described nightmarish dreams he often had – going through dark tunnels with terrific apprehension, or of being chased with feet as heavy as lead; and a medical student mentioned a constant nightmare of 'something – a bat or spider – seizing hold of my tummy which gave me such excruciating pain that I cried out, and I woke up crying.' Finally, a quiet young woman asked about her dreams replied: 'I'd really rather not say!'

Thus men and women have talked about their dreams for centuries and centuries, and the curious thing is that in spite of the bizarre and nonsensical nature of dreams, people have persisted in believing that dreams have got a meaning and that they are significant for our daily lives.

Dreams are of interest to all of us, because all of us dream. Some find them amusing, some worrying, some terrifying, some regard them superstitiously, gypsies take them as warnings, the superior regard them as merely stupid, while others, like the psychoanalysts, study them scientifically and make use of them in the everyday treatment of neurotic disorders. Dreams have always fascinated mankind, because they open up to us a realm of our personality which we did not know existed. We are surprised at the things we dream about. St Augustine thanked God he was not responsible for his dreams! The mild person dreams of himself standing up to others, the humble person of great achievements, the philosopher dreams of doing the most irrational things, the calm and collected individual of being in a berserk rage. Dreams indeed seem to go by opposites. What does it all mean? Is it all chance, is it all nonsense, or has it all a meaning for our lives? At any rate, we cannot help being fascinated by the bizarre products we find emerging from the depths of our personality, and which are the products of our own unconscious creation.

For night after night the most unimaginative of us go on creating images and stories so fantastic that we could never produce them by any conscious effort in our waking life however hard we were to try. Not only do we create these stories, but we are ourselves partakers and actors in this phantasmagoria we unconsciously fabricate; and reluctant as we are to admit it, we are affected by it in our daily life, and are often, like Pilate's wife, troubled by them.

The child who has had a vivid dream of being bitten by a dog will be afraid of dogs the rest of the day – the dream stands for reality. If he dreams of his mother as an avenging witch, he will be suspicious even of his mother's affectionate advances during the day, lest, as in the story of Red Riding Hood, the kindly mother turns out to be a devouring wolf. Who of us can say that we have not been affected by some depressing dream, which has left a marked effect on the life of the day? Such dreams cast their shadows upon the whole day. A humiliating dream will make us subdued through the day; terrifying nightmares will make us apprehensive, whereas a frustrating dream will have the effect of making us irritable all morning. But dreams also have a positive effect on our lives so that the aggressive dream of the timid man in which he stands up against his superiors will cause him to be more assertive, and dreams of success will cheer us up all day. We cannot therefore afford to ignore dreams, for however much we may scoff at them as nonsense, they have an undoubted influence upon our daily life.

THEORIES THROUGH THE AGES

Primitive man believed in them most of all: for believing as he did that the soul was detachable from the body in sleep, he considered that if in his dream he saw himself in a neighbouring hostile village, he had actually been in that village, and if he saw the villagers in his dream preparing for battle, this would be quite enough for him to report the fact, and his tribe would immediately prepare to meet the onslaught. Lévy-Bruhl[1] says, 'The outlook of the Red Indian is a very practical one. He believes that man has two souls, one simply the vital principle of the body which dies with the body, the other dwells in the body but leaves it at death. This soul is his guardian angel, his genius, his protector, his personal god, his genius upon whom he depends. He is therefore responsible for what his soul does in dreams.'

So strong is this conviction of the reality of what is experienced in dreams, that when an American Indian dreamed that the missionary had stolen his pumpkins he openly accused the missionary of doing so, and persisted in the accusation although it was pointed out to him that at the time of the supposed theft the missionary was 200 miles away, and that in point of fact the pumpkins were still there in his plantation. But he still demanded compensation, because he said the missionary would have stolen them if he had been here! So persistently did he believe the evidence of his dream against all evidence both of reason and of the senses!

Thus primitive man believes in his dreams even more than he does in his own perception, and he relies upon them for guidance in everyday affairs. If a savage dreams that he owns someone else's property and tells the owner, the latter will say to him, 'Take it, it is yours!' Not only are people held responsible for what others see them doing in their dreams, but the accused man is ready to accept responsibility for what the other has dreamt he was doing. If a man dreams of making love to another man's wife and it becomes known, he is taken to have done so, is punished, and accepts the guilt and punishment as merited – perhaps not without reason! The modern instance is that of a young wife who dreamt that her husband was making love to a blonde and

1. *Primitive Mentality*.

was furious. Being reminded that it was only a dream, she replied, 'Yes! but if he does that sort of thing in *my* dream, what will he do in his own!'

In what we might call the middle period, between the primitive and the classical, we have the dream of Pharaoh about the seven fat kine being devoured by the seven lean kine, interpreted by Joseph as referring to seven years of plenty being followed by seven years of famine. Acting upon this, granaries were set up, the corn stored during the years of plenty, and the famine avoided. Joseph was therefore raised to great power in the land. But this only confirmed a prophecy in an earlier dream of Joseph in which his brothers' sheaves bowed down to his sheaf, meaning that his older brothers would bow down to him – a dream the brothers took so seriously that, intuitively recognizing the meaning of the dream, they put him in a pit to prevent the dream coming true. But the dream did come true, for in the years of famine the brothers came and begged corn of Joseph, whom they did not recognize to be their long-lost and 'murdered' brother. Such dreams were obviously regarded as prophetic and premonitory.

DREAMS IN CLASSICAL TIMES

In classical times dreams were the means by which the gods made their communications with man. For instance, in Homer's *Odyssey* the goddess Athene came in a dream to the girl Nausicaa – who for naturalness, modesty, and tact is surely one of the most charming characters in literature – and in the form of one of Nausicaa's friends the goddess told her it was time she gave up her slovenly ways and began to wash her linen, which was piling up, for 'that kind of thing gives a girl a good name in the town. ... Look at the lovely clothing you allow to lie about neglected, although you may soon be married and stand in need of beautiful clothes.' She urged her in the dream, therefore, to collect all her clothes next day and go with her maids for a picnic to cleanse them in the washing-pools – a striking picture of a girl's intuitive awakening from hoydenish adolescence to maidenly attraction. It was at the washing-pools that she met the storm-tossed Odysseus – 'the kind of man I should fancy for a husband' – as the wily goddess intended.[1] Again, when the

1. *The Odyssey*. Trans. by E. V. Rieu (Penguin Classics).

Greek Army before Troy was smitten by a pestilence, Achilles suggested that guidance might be obtained by means of dreams, 'because dreams descend from Zeus'. Many natives in modern times only regarded themselves as converted to Christianity when they *dreamt* that they were converted, and not before, for dreams were the voice of God. St Paul, we remember, was converted by seeing such a vision.

MODERN POPULAR THEORIES

In modern times there are many theories of dreams, some merely popular, some bizarre, some more scientific; but even the scientific theories differ widely according to different Schools of psychopathology.

A consideration of these theories of dreams which we shall now study will, I think, convince us that there is considerable truth in all of them, though none of them is adequate as a complete explanation. This applies both to the popular views which we consider in this chapter and to the theories of the accepted modern schools which we shall study in later chapters.

(a) *The Physiological or Heavy Supper Theory*

There are some who believe in the physiological or heavy supper theory, namely that dreams are caused by indigestion and other bodily disturbances. There is probably some truth in these theories. A heavy supper, by drawing the blood for the digestion, may affect the circulation to the brain and so give rise to the dream. But whilst this may account for the fact *that* we dream, it does not account for *what* we dream. Four men have a supper of pork and beans, but whereas one may dream of his lady-love, another dreams of being pursued by horrible monsters, another of being murdered, another of failing in business. There is nothing in the chemical composition of the pork and beans to account for these differences. It is more probable that they already had these subjects upon their minds and that the indigestion merely precipitated them into the form of a dream. In other cases, however, as in the case of the sailor cast adrift who has the hallucination of drinking lakes of water, the content of the dream as well as the fact of the dream is probably determined by the physiological state of thirst. The same applies to sex dreams, which are often precipitated by our physiological state of tumescence at the time, so that not only the precipitation but the subject-matter of the dream is at least partially due to such physiological tension. But not entirely, for whilst our physiological state may make us dream of sex, the specific nature of the sex dream, whether natural, homosexual, or fetishistic, has yet to be explained. Another physiological theory, which has been advanced to explain the more specific details of the dream, is that dreams as well as the hallucinations arising from the delirium of fever are simply caused by the irritation of certain cells of the brain, so that the memory traces contained in these cells are thrust into consciousness. This is a very reasonable theory, which would account for *what* we dream about, as well as for the fact *that* we dream.

Strongly in favour of the physiological origin of dreams is the fact that they can be experimentally produced. If one stimulates an area of the exposed human brain and tickles it with an electrode, the stimulation sometimes causes the individual to dream, even while he is completely aware of what is going on around, and afterwards he can describe his dream. For instance, 'I was in my mother's house. My sisters and mother were there. Everyone was talking.'[1] If a dream came externally from the stimulus of an electrode there is no reason why it should not come internally from irritants in the blood stream.

Such a theory would explain also the bizarre and kaleidoscopic nature of our dreams and why bits of old memories or fragments of past experiences are revived. According to this view, the dream is purely accidental, depending on the particular spot in the brain which happens to be irritated, and has no psychological significance or meaning. But there is an objection even to this attractive theory. For one thing, dreams are rarely pure memories as they would be if they were mechanical reproductions of past events; but there are almost always modifications and changes even of the simplest experience of the day. Apart from that, presumably in a fever the *whole* of the brain is washed by the blood stream containing the toxin. Why then should these particular cells of the brain be selected and aroused into dream consciousness and not others?

This can be explained if we assume that it is those cells of the brain which are *already in a state of emotional tension* which are

1. William Penfield, *The Physical Basis of Mind*, chap. vii, p. 61.

thus stimulated into activity by the toxin to produce the dream; whereas those parts of the brain which are in a state of equilibrium are not so affected, although subject to the same toxin. Thus it is natural to suppose that if we have had some experience in the past like that of fear, the emotion of which has remained undischarged, or some anger that we have had to suppress, the cells of the brain which retain these bottled-up emotional experiences will remain in a state of tension, and will be the ones most readily aroused by the toxic process and give rise to the dream or hallucination. In that case the dream (even if we suppose it is stimulated by some physiological irritant) is not purely accidental, but is determined by the potential energy contained in certain cells of the brain in a state of tension as the result of some undercharged or repressed experience of the past. This is a very tenable theory, but in suggesting this explanation we have stepped over into the modern *psychological* explanation of dreams as due to repressed emotion producing the state of tension. The pork and beans, to return to our earlier illustration, explained that the gentlemen dreamt; but the fact that one dreams about his lady-love, another about murder, another about failure, probably points to repressed emotional experiences with which their minds were already preoccupied. Evidence of this is to be found in the fact that when we analyse out even the hallucinations of the fevers we find that they relate to such experiences. For instance, a boy of eight in a high fever had terrifying hallucinations of being pitchforked into hell and burning in its flames. The heat of the fever made him dream of being consumed by flames, but that was not all. Nor was the dream merely due to an accidental stimulus of those cells of his brain which contained the memory of what he had heard about hell, but it related to the fact that it was because he had been disobedient to his mother and gone out without a coat that he had caught pneumonia. He thought he was going to die, and as naughty boys went to hell, pictures of which he had seen in the family Bible, the horror assumed the form of this nightmarish hallucination. Not the toxin alone, but the emotional problem which was worrying him, created the dream. If the dream was caused by the irritation of cells of the brain, it was those cells which were already in a state of emotional tension which were particularly affected and produced the specific dream.

The same principle applies to those dreams which are directly due to *external stimuli*, of which there is ample evidence. The ringing of an alarm clock instead of waking up the dreamer, made him dream of church bells and of sleighing in Russia. Such dreams are common experiences of all of us. Recently I went out shooting before breakfast. At the first shot, my son, who was asleep, dreamt that it was the first of a twenty-one gun salute to Princess Alice and went on sleeping. Why Princess Alice I did not inquire, but the 'twenty-one' enabled him to anticipate more shots and therefore to continue to sleep in spite of them. A friend of mine dreamt so vividly that the house was on fire that he got up to investigate, only to find that the gas had been left on unlit in the basement three storeys down. The smell of this, even so far away, made him sense danger. Such dreams are obviously due to external stimuli.

But such an explanation carries us only so far. Our minds are constantly subjected to external stimuli even in sleep; why did these particular stimuli produce dreams when numerous others passed unnoticed? The answer is that in each case the stimulus aroused some problem in the mind – producing a state of tension. This would certainly apply to the dream of the fire from the smell of escaping gas, which would make the dreamer subconsciously sense danger, and which in fact made him get up in alarm. In the case of the alarm clock and the gun, the corresponding dream arose from the problem of the individual's reluctance to get up. If these dreams had been due to the tickling up by toxins of certain cells in the brain which contained memory traces of earlier experiences, the result would have been an exact reproduction of these experiences; they would have related simply to gas, clocks, and shot-guns. But they did not: for them were deliberately substituted sleigh bells for the alarm clock, and salute guns instead of actual guns, designed to evade the inconvenient issue of having to wake up, to relieve the state of tension, and so allow the dreamer to sleep.

There are other dreams which are often attributed to external stimuli but which probably have a deeper significance. The nightmares of passing through a long narrow tunnel used to be explained as due to the subconscious sensation of the passage of food through the intestinal passage, but in all such nightmares as we have investigated the terror related to a birth trauma, the hazardous passage through the uterine tunnel with the suffocation and dread associated with it. This terrifying experience can leave its traces behind and is a common cause of claustrophobia as it is of other anxiety states.

This leads us to dreams arising from *internal* stimuli. Aristotle originated a theory which Jung and others have more recently elaborated, that dreams 'call attention to incipient morbid conditions of the body which have escaped the notice of our waking state'. Indeed organic diseases may sometimes be diagnosed from such dreams since the sensations arising from diseased organs may be subconsciously perceived in dreams though not in waking consciousness. A dream of this kind is given in our description of precognitive dreams later in this chapter. A man dreams repeatedly that he has cancer of the tongue and develops a phobia of it long before there is any indication of it whatever: nevertheless he developed this disease a year later. Was this pure chance; or did the fear of the cancer bring it about; or were the dreams and the phobia due to the fact that the cancerous processes had already started unperceived except subconsciously and also in his dreams? All these theories are possible, but we think the last most probable.

(b) *The Personal Reminiscence Theory*

This popular and commonly accepted theory is that we dream of what happened to us the day before, or years before; that dreams are simply reproductive of past events. There is obviously much truth in this theory, but it requires modification. For one thing, when we examine such a dream we find that though it is a reproduction it is rarely an *exact* reproduction of what happened; almost invariably it differs from the experience of the day either in details of fact or in emotional tone. This is a fact, as we have pointed out, which the physiological stimulus theory is also inadequate to explain. The man who is nagged by his wife and takes it lying down dreams that he strangles her – evidently not a pure reproduction, but it is what he would like to have done but equally evidently what he was afraid to do. We dream of walking down the High Street naked, but instead of being overcome with shame and confusion as we should be in fact, we take it as a matter of course; or if we are a bit embarrassed, nobody else seems to take any notice.

In any case, on the purely reproductive theory, why do we dream about these special experiences out of the thousands of things we have done, seen, and read, and these not by any means the most important, nor necessarily those charged with most emotion? They are often concerned with the most trifling experiences. The reason seems to be that these persons or situations carry special significance for our lives, symbolic and otherwise. For instance, the person we casually met or the apparently insignificant event of years ago come to mind in dreams because they touch upon some specific problem of our lives. Recall the case of the woman who dreamt of Largs, a place in Scotland which she had visited twenty years before and then only for a night. We may say that the dream was purely accidental, due to some chance association, or due to the fact that her brain was stimulated in those precise cells in which the memory of this town resided. But there was a more specific reason why this particular experience was selected for reproduction; from an investigation it became clear that on that particular night twenty years before she had had an argument with her father to try and persuade him to allow his other daughter to marry a man of whom he disapproved. But even so, why should she recall that experience so long after in a dream? It was because she also had later married against her father's wishes, and now her own marriage was turning out a failure. The emotional situation was similar, but though apparently concerned with the remote past was really related to a present personal problem. But why did it appear in a dream? Because it was a problem which worried her but which she was reluctant to look at and face in the day. By recalling the situation at Largs, the dream is reminding her that she had only herself to blame, a fact she was conveniently overlooking. The dream simply says, 'What about Largs? You remember what happened there! You, like your sister, went against your father's advice and you must face the consequences.' The dream, in other words, was not merely reproducing a situation; it was recalling a problem, which is an important distinction.

(c) *The Theory of Racial Reminiscence*

While there are those who regard all dreams as personal reminiscences, there are others who explain many dreams as reverberations of the experiences of our remote ancestry, as atavistic, as racial. The dream of falling, just before dropping off to sleep (it is significant that we speak of 'dropping off'), may be rationally explained physiologically as due to the sudden lowering of blood pressure: this is probably frequently the case. Or it may be a personal reminiscence of a fall in childhood which produced a conditioned fear of falling; or it may be explained as a fear of a moral lapse (we speak of a 'fallen woman') – a symbolic explanation. But some explain these dreams of falling as atavistic, as an echo from the time when we lived in trees, when falling was of course a common danger.

This explanation is only partly true. It is probable that we inherit a predisposition to certain fears, like fears of falling or the fear of suffocation, and such fears are likely to emerge in dreams. But why then do only some people dream of those things? It suggests that whilst the predisposition is there in all of us, the precipitation comes from some more personal experience in childhood – even though this experience might assume a moral significance later. Many cases of claustrophobia can be traced back to an actual suffocation at birth and so cured; but in many cases this claustrophobia relates to a man who has a power psychology and cannot bear to be frustrated, thwarted, shut in.

Jung has given a new significance to the theory of racial reminiscence by pointing out that we dream of many things which we have never personally experienced, and for an explanation we must therefore go back to our more remote ancestry. This theory we shall describe later as the archetypal theory.

(d) *The Premonitory Theory*

The theory that dreams foretell the future, or that 'dreams come true', has always been widely accepted. Primitive man believed in it firmly and absolutely. Pharaoh's dream of the fat and lean kine and Joseph's dreams of the sheaves and stars were regarded as premonitory; and many people at the present day claim to have dreams foretelling success, death, or disaster. To accept this theory in its entirety we must of course not only believe that the future is entirely fixed, but that there is a possibility of our knowing it, both of which are vast philosophical presuppositions though not necessarily untrue. An exposition of this theory is to be found in Dunne's *An Experiment with Time*, in which he not only finds evidence of anticipation of future events in dreams but also suggests how such premonition can take place. He states that if we keep a record of our dreams over a period of months we shall find a large percentage of them fulfilled by the events. To discuss this theory is beyond the scope of our study. It is only right, however, to point out that many scientists, even some physicists, now believe in the possibility of premonition; and some of the facts of the case as well as Saltmarsh's theory to explain them are cited in our chapter on 'Dreams and Psychical Research' (Chap. 11).

But whether or not there is objective truth in this theory, there is undoubtedly *psychological truth* in premonition, since we may intuitively and subconsciously foresee the future to which we may be consciously blind. We may in a dream see a man we trust doing the dirty on us, and that may be because although we tell ourselves that he is all right, our intuition senses him as a twister. We define intuition as subconscious inference, that is to say, inferences drawn from slight indications which we do not consciously perceive. Such inferences may be perfectly justifiable and may give a more accurate judgement of men and events than our conscious observation and reason. The dream, which reveals the workings of the subconscious mind, may thus give us insight into the character of others or make us aware of the trend of events which thereafter comes true. This, after all, is not such an unusual process; some people can tell you that it is going to be a fine day without being able to tell you why, or how they know; but it is from slight indications registered from past experience and now subconsciously inferred. Possibly Joseph's experience as a stranger in Egypt had led him to perceive over the years what the natives had failed to see – the cycle of the years and of good and bad harvests.

Similarly, intuition, and therefore dreams, may give us warnings of what is taking place in ourselves. A man in business full of self-confidence dreams that he fails in business and bluffs himself that of course dreams always go by opposites. But later he does fail, although he did not expect it. His dream may be regarded as premonitory; so it is, but only in the sense that his intuition in the dream was warning him that he was too cocksure of himself (which his friends could also have told him) and blinded to his own weakness. Indeed his over-self-confidence was a disguise to hide his inferiority and over-compensation; and it is this sense of inferiority together with the compensating conceit which brings about his downfall. The dream revealed his true nature to himself, and therefore suggested that if he continued in that cocksure attitude he would come a cropper, as he ultimately did. In fact he crashed, not because he dreamed it, but he dreamed it because his intuition told him that such over-confidence would lead to disaster. The dream was premonitory in the sense that it was a warning; but it was premonitory only in a psychological, not an objective sense. On the other hand, the dream of Joseph that his brother's sheaves bowed down to his sheaf, which came true when he rose to high position in Egypt, was an expression of his early subconscious ambitions of greatness which his inferior position in the family would not permit him openly to express. That dream also was premonitory, but the fulfilment came not from the dream, nor from the objective anticipation of events, but from his latent ambition. The dreams of organic illness previously mentioned provide a still more striking illustration of what appears to be premonitory, but which in fact is subconscious inference from present sensations. The most striking case of this kind we have met was of a patient who dreamed on several occasions that he had an attack of paralysis of his arm and mouth which went into a state of spasm. Some months later he did in fact have exactly such an attack whilst mending his radio and he did become partially paralysed. Was his dream 'premonitory' of this attack? No! It was due to an arterial spasm caused by congenital syphilis. No doubt on the earlier occasions he had had slight attacks in his sleep which gave rise to the dream, but later he had a more serious arterial spasm when awake which left him more permanently paralysed until he was successfully treated for his disease.

VIEWS OF PHILOSOPHERS AND EARLY PSYCHOLOGISTS

Dreams have naturally attracted the attention of philosophers, and of psychologists when psychology began to be an independent science. The topics with which these more systematic thinkers have concerned themselves have been in the main four in number: (i) 'The stuff that dreams are made of'; (ii) the forms which they assume; (iii) their 'efficient' or exciting causes; and

(iv) their psychological or biological function. This fourth aspect of the theory of dreams, however, becomes prominent only in the most recent and contemporary theories.

Concerning the 'material' of dreams, it has been very commonly held that they are composed of 'mental images' – mainly visual; but there has been a strong and persistent tendency to believe that every dream has a sensory nucleus, some actual sensation arising from stimulation of the dreamer's sensory organs. The problem of 'form' that has been most discussed is really the problem of the apparent *formlessness* of the dream, the apparent irrationality and incoherence. Before Freud, no one attempted seriously to develop the idea that a dream has a form in the sense in which we speak of the form or structure of a play or a novel, and before Freud no one attempted to describe the 'mechanisms' through which form in this sense is produced.

The earlier attempts to give a causal explanation of the dream were in terms of the effect of chance stimulation of the sense organs and the chance-like operation of the 'laws of association'. The nearest approach to a functional explanation of the dream was to say that in the dream the dreamer is trying to interpret or make sense of vaguely experienced sensations; and he does this by supplementing the sensations by associations just as we do in the interpretations of waking life. The influence of sensory stimuli on dreams has already been referred to. Thus a sleeping person, vaguely conscious of the hot-water bottle at his feet, may dream that he is climbing Mount Etna and walking over hot lava.[1] In all these early theories there is an almost complete failure to assign any influence to motives other than the purely intellectual desire to make some sort of sense out of meaningless

1. The laws of association so invoked were generally classed as Association by Contiguity, as when lightning suggests the idea of thunder, and Association by Similarity, as when one person's features make us think of another person whom he resembles. Most dreams were explained in terms of association by contiguity, but examples are also cited of dreams determined by association by similarity. Thus Maury, a French psychologist, reports that on one occasion he dreamed that he made a pilgrimage (*pélerinage*) to Jerusalem where he met the chemist Pelletier who gave him a shovel (*pelle*). On another occasion he dreamed about *kilometres*, *kilogrammes*, the island *Gilolo*, the flower *lobelia*, General *López*, and a party of *loto*. Another illustration is given on page 18, in which *potato* is substituted for *peut-etre*.

stimuli. Freud's theory was revolutionary in discerning motives much more powerful, meanings much more important, and in showing that a dream had a function or purpose. The transition from a mechanical-associationist explanation to a functional explanation of dreams was not, however, quite abrupt. A transitional phase can be found in the doctrines of the German psychologist Herbart. A distinctive feature of Herbart's general psychology was the conception of ideas as 'forces' – the doctrine that every idea actively strives to gain entry to consciousness, that mental life is, so to speak, a continuous battle of ideas each of which is struggling for the limelight of man's conscious attention. This entails a concept of repression, since the ideas that lost the battle are banished into the darkness of the unconscious, where they may still continue to organize themselves as a sort of underground movement preparing for the next assault upon the conscious mind.

Herbart's influence was very strong at the end of the nineteenth century. His views were very much in the air and there can be no doubt that they influenced Freud. They also influenced the French philosopher Henri Bergson, whose theory of dreams is the most systematic and coherent of all the pre-Freudian theories.[1]

Bergson was one of those who believed that every dream has some actual sensation as its core and its provocation. But this nuclear sensation is vague, indeterminate, and confused. Such impressions constitute the raw material of the dream, but they cannot produce it alone. Something more is needed to determine the dream's specific form. This something more is the memory and the memory alone. Bergson was also one of those who believed that every memory is indestructible, that the whole of a man's past life is preserved in his memory, even to the most infinitesimal detail. What is more important, all these memories are (as Herbart's doctrine implied) dynamic forces. Each struggles to regain the light of conscious recollection. During waking life they are repressed by waking interests, but sleep is essentially a state of *disinterestedness*. Sleep thus unlocks the trap-door in the floor of consciousness, and memories arise from

1. Bergson's theory was first presented in a lecture before the Institut Psychologique on 26 March 1901. It was published in the *Revue Scientifique* of 8 June 1901. An English translation revised by the author was published in book form by T. Fisher Unwin in 1914.

the depth to perform 'in the night of unconsciousness a great danse macabre'. Even so, although all memories clamour for admission, only a few get through, namely those that most readily combine with some present sensation. When the union is effected between this present sensation and the most congruous memories we have a dream.

A consideration of these popular and philosophic theories of dreams will, I think, permit us to conclude that there is considerable truth in all of them, though none of them, as we have seen, is adequate as a complete explanation. This applies also to the more scientific theories of the accepted schools, which we shall later describe.

DO WE DREAM ALL THROUGH SLEEP?

Meanwhile we may ask the question constantly put forward, 'Do we dream all through sleep, or only, as some think, just before waking?' Opinions differ. It is possible that we may be in a state of such profound sleep that consciousness is completely abolished, and that we dream only when we are waking. Or it may be that we are dreaming all through sleep, but *remember* only what we dream just before waking, for it stands to reason that the nearer to waking we are, the less dissociated is the dream, and therefore the more easily remembered. In that case, to say we do not dream simply means that we do not remember having dreamt.

Whether we are dreaming all through sleep or not, it is probable that the mind is far more active throughout sleep than we imagine; and certain it is that we dream far more than we normally remember. How often some trifling event of the day will make us remember a dream which would otherwise be buried in oblivion. Further evidence of this is that people talk in their sleep and yet continue to sleep. They must be dreaming to be talking, and yet this is not 'just before waking', for they neither wake nor do they remember anything about it on waking. Indeed they get a shock when their wives tell them that they have talked in their sleep, and are still more embarrassed – or it may be relieved – when they hear what they have said in their sleep! Again, if a patient is hypnotized and left for a time in sleep, and is suddenly asked what he is dreaming about, he will be able to give an account of his dream. All this suggests that the mind has been actively engaged all through sleep in worrying out problems, even though we may afterwards be entirely unaware of what has been taking place.

Further evidence suggestive of this conclusion is that the brain is known to be in a state of activity throughout sleep, for it has been demonstrated that brain waves of a certain type and rhythm are given off all through sleep, as measured by the electroencephalograph. Of course activity of the brain does not necessarily mean activity of the mind, although Braizier has shown by such experiments that 'dreams are associated with bursts of electrical activity'.

It is perhaps safest to conclude that there are brain processes which are not accompanied by any kind of awareness, that others are subconscious and can be remembered only with difficulty, whereas in other cases the brain activity is of such intensity that it is accompanied by vivid dreams which are easily remembered, and indeed not easily forgotten. As to the question whether we dream only on waking, we suggest that instead of saying that we dream only just when we are waking, we should say that dreams tend to wake us. In other words, it is not the waking which produces the dream, but the vivid dream which often produces the waking.

Dreams have been described as 'the activity of the mind during sleep'. They are also described as 'the language of the unconscious'. Both of these are good enough as descriptions, but they are inaccurate as definitions. Indeed neither is strictly true; for dreams sometimes deal with material and problems of which we are perfectly conscious, but which we are trying to shelve, and are not then 'the language of the unconscious'. And there are dreams which are specifically *day* dreams, and not due to the activity of the mind during sleep only. There must be some characteristic which links up the dreams of the night and day-dreams and makes us call both of them 'dreams'. Perhaps the broadest generalization is that the characteristic of dream activity is that it takes place irrespective of any conscious effort or direction on our part, which condition naturally occurs more often when we are asleep than when we are awake. This is characteristic of all dreaming, and gives rise to primitive man's belief that dreams are due to some outside agency, and to some modern scientific views that they are the activity of an entity called the 'unconscious mind'.

CHAPTER 2

DREAMS AS WISH-FULFILMENT

(1) FREUD'S THEORY: INFANTILE SEXUAL WISHES

THE theory of wish-fulfilment, associated with the name of Freud, contains much obvious truth. A Welsh tenor singer dreamt that he was singing in the Heavenly Choir conducted by the King of Heaven. The choir consisted of a thousand sopranos, a thousand altos, a thousand basses, and *one* tenor – himself. Then the Great Conductor struck up and the whole choir started the chorus; but after a few bars He tapped His music-stand and, turning to the tenor, said, 'Tone it down a bit, Mr Jones, tone it down!' The wish-fulfilment is not far to seek in such a dream. So Arctic explorers dream of sumptuous meals, the ascetic hermit dreams of enticing girls, the boy of heroic deeds, the girl of being a princess, the clerk of being head of the firm. In this respect they resemble day-dreams. The Hungarian proverb says, 'The pig dreams of acorns, the goose of maize' (quoted by Ferenczi).

But Freud goes further: he says all dreams are not merely wish-fulfilment but fulfilment of *sexual* wishes. Dreams are the language of the unconscious: and since, according to Freud, the unconscious consists entirely of sexual wishes, dreams relate to such repressed wishes.

The sexual nature of the dream in many cases is obvious and undisguised, as in the case of the hermit mentioned or the youth whose erotic dreams represent his unsatisfied desires. Sometimes they are symbolic but only slightly disguised. Thus the dream of the young woman that she is being stabbed in the stomach with a bayonet which has a busby at its butt end is obviously of a sexual nature. In most dreams, however, the sexual element is completely disguised and the dream appears to the layman to have no connexion with sex at all; but in all of them Freud finds a sexual meaning notwithstanding. Indeed he holds that 'Dreams which are conspicuously innocent invariably embody coarse erotic wishes.'[1]

A typical case of a dream not manifestly sexual, but symboliz-

1. *The Interpretation of Dreams*, p. 241. (When not otherwise stated, the pages in this chapter refer to the 1913 translation by A. A. Brill.)

ing sexual desire, is the following, as given by Freud: 'Between two stately palaces stands a little house receding somewhat, whose doors are closed. My wife leads me a little way along the street up to the little house and pushes in the door, and then I step quietly and easily into the interior of a courtyard that slants obliquely upwards.' Freud interprets this as a desire for sexual intercourse from behind, the buttocks being the two stately palaces, the narrow slanting passage the vagina; whilst the little house reminds the dreamer of a place in Prague, and thus of a girl whom he had met the day before who came from Prague, and with whom he would have liked such intercourse.

Freud concludes that 'the more one is occupied with the solution of dreams, the more willing one must become to acknowledge that the majority of dreams of adults treat of sexual material and give expression to erotic wishes.'[1]

These sexual wishes from the unconscious, says Freud, have to be repressed. This creates a tension, and this instinct tension, denied outlet during the day, discharges itself at night in imagined gratification as dreams, and this allows the individual to sleep.

The reason for the repression of the sexual wishes is partly because of the demands of life. We cannot for ever indulge in wishes, for we have to live and meet the needs and difficulties of reality; and the wishes of the unconscious being contrary to those needs and demands, the pleasure principle conflicting with the reality principle, they have to be repressed. Another reason for repression is that for a boy, for instance, to wish to possess his mother sexually meets with the hostility of his father and the threat of castration at his hands. It is therefore the function of another part of his personality which Freud calls the Preconscious to guard the person against these dangerous sexual wishes from the unconscious and keep them in check.

But these desires are too strong to be completely repressed, and later emerge, sometimes as neuroses, such as anxiety states which express the fear of these forbidden desires, and obsessional acts by which it is hoped to propitiate for them; or they may emerge as dreams. Dreams are therefore the emergence into consciousness of repressed unconscious sexual desires.

But when these forbidden wishes appear in dreams they must,

1. *The Interpretation of Dreams*, p. 240.

as a rule, still be disguised to evade the censorship of the preconscious which is ever on the watch to keep them from emerging into consciousness.

There are many forms of disguise and many means of evasion, a common one being the use of symbols – much as the comedian often evades the censor by indulging in a *double entendre.* (I don't mean what you mean, says he when the audience laughs at such a joke.) So symbols are capable of a double meaning. A bayonet is itself a dangerous instrument and the girl may easily bluff herself that that is all she feared; but it also represents a phallic symbol, the male sex organ, which in the dream is the real cause of fear – the fear of a desire. The symbol has a double meaning, and where the censor would not allow direct reference to sex either on the stage or in our dreams, it allows this apparently harmless reference to the bayonet. The same applies to the symbolism of the little house with the closed door between the two stately palaces. The boy might wish to kill his father to possess his mother; but that is not only forbidden, it is also dangerous; therefore he dreams of an assassin killing the king, thereby evading the guilt, but gratifying himself with the thought of assassination.

Freud has thus worked out many of these symbols and their corresponding meanings, all of which of course have reference to sex: a snake, a dagger, or anything long represents the male sex organ; anything like a box, a room, or anything which contains things represents the female uterus; a king represents the father; a queen, the mother (as sexually desired), and so on.

If you once accept the completely sexual nature of the dream, it requires very little ingenuity so to interpret dream symbols. Thus if you dream of a piano: piano...scales...stairs...the rhythm of the sexual act.

Thus Freud says, 'There is no series of associations which cannot be adapted to the representation of sexual facts.' This remark may be interpreted in a more cynical sense than Freud intended – that once you accept the sexual theory the interpretation may easily be adapted to correspond with the theory – which is perhaps nearer the truth!

Freud's theory has often been criticized on the grounds that our dreams are as often as not about simple events of the day; and, further, that they apparently have nothing to do with wish-fulfilments, sexual or otherwise: indeed most of them are unpleasant.

That dream material is commonly derived from events of the day is admitted by Freud, but he maintains that even such dreams have a hidden meaning which if explored reveals a secret sexual wish. For he distinguishes between the *manifest* content of a dream which provides the material for the dream picture, and the *latent* content of the dream which is the dream's real meaning and significance. The dream, like the parable, is a symbol taken from everyday life to give expression to a more subtle and hidden meaning. Maurice Nicoll[1] gives the best explanation of this distinction between the manifest and latent content of the dream when he likens the dream to a cartoon. In the evening paper you may see the cartoon of a bus fallen down in a street under repair and you may say, 'I know the explanation of that cartoon, for such an accident happened in Fleet Street this morning and the cartoonist no doubt saw it.' That explains the manifest content which suggested the picture; but it does not of course explain the cartoon, which is meant to represent, shall we say, the way in which the world is in danger of crashing over some international dispute; and to make it clear the cartoonist may label the bus 'The Nations' and the hole in the road 'The Eastern Crisis'. That is the latent content, the real meaning and significance which the cartoonist wished to convey. So we may superficially explain a dream merely by saying we dreamt it because of something which happened during the day; but there is usually some deeper reason why that incident or experience is chosen out of all others of the day as the subject of the dream. For in itself the incident may be of little importance, a passing glimpse of someone in the street, a passing reference in the paper, but to the dreamer it recalled something which has significance for him personally.

A man went to see the play *Hamlet* and afterwards had a troubled nightmare about it. Anyone might think that the dream was fully explained by the fact that he had been to the play, which upset him. But why should it upset him and produce such violent emotion? The true significance of the dream only came when he realized that Hamlet was unconsciously in love with his mother, and therefore wished the death of his father, his rival:

1. *Dream Psychology*, Chap. iv.

and that was why Hamlet was so impotent to bring himself to avenge his father's death. The play unconsciously aroused similar complexes lying latent in the dreamer's unconscious mind and gave him an uncomfortable night. The event of the day before explains why he dreamt of the play; that is the manifest content: it does not explain why he had such a disturbed night, which is only explained by the latent meaning, of the play as well as of the dream, which was his own infantile Oedipus complex, the sexual desire towards his mother, and the wish, therefore, to be rid of his father.

As to the purpose of the dream: we have seen that the conflict between the forbidden sexual wishes and their repression by fear of consequences produces a tension in the personality, which, denied outlet in the day, discharges itself in dreams. The function and purpose of the dream is therefore, according to Freud, to relieve that tension by gratifying the wish, and so allow the individual to sleep. 'Dreams are the means of removing by hallucinatory satisfaction mental stimuli that disturb sleep.' 'The Dream', says Ernest Jones, 'is thus the guardian of sleep.'

But if the function of dreams is to gratify repressed wishes and allow us to sleep, it obviously often fails of its object, as shown by the illustrations already given, for the majority of dreams are unpleasant, and many are obviously associated with anxiety and dread. This is especially true of nightmares. We may dream of being found guilty, condemned, and punished; or we may dream of being chased, overwhelmed, and of retribution overtaking us. They are certainly not gratifying; they certainly do not allow us to sleep, but often wake us so that we fear to sleep.

Freud again explains this by saying that this unpleasantness is the work of the inhibiting forces of the pre-conscious, whose job it is to restrain the pleasure principle of the unconscious from invading consciousness. There are in fact two systems, and what is pleasurable to one is painful to the other. These primitive sexual feelings, in other words, are an intrinsic danger, they deserve punishment, and it is that which is reflected in the dream. The girl with the bayonet dream wakened up in terror, which not only implies the strength of the sexual wish, but a fear of this wish in her, possibly on account of a sense of guilt. *The purpose* of the dream was nevertheless to give expression to the sexual wish – according to Freud's view. But the fact remains that it is the unpleasant aspect and not the pleasant wish which is dominant in this as in most other dreams.

The two outstanding contributions of Freud to the study of dreams are, first, the conception of dreams as having a function and purpose (although we may not agree that the whole purpose is the fulfilment of a sexual wish); and, secondly, his analysis of what he calls the 'dream work'.

THE MECHANISM OF DREAMS

One of the greatest contributions Freud has made to the theory of dreams is what he calls the 'dream work', or the mechanism of dreams.[1] This is made necessary by the fact that the dream is trying to express wishes of the unconscious which are incompatible with the self and therefore must be transformed into a form approved by the self. This is primarily the work of the preconscious, whose sphere of operation lies between the unconscious and the conscious. Without an understanding of those mechanisms it is almost impossible to understand the meaning of dreams, for dreams otherwise appear quite distorted.

Displacement is one means by which the dream does this. In the first place, it is the effect of feeling which is displaced, so that what appears to be most important and full of emotion in the dream may be really of least significance, whereas an apparently trifling matter may be pregnant with meaning. The emphasis of the dream is thus displaced and so eludes the censor. Similarly there may be a displacement of persons, so that a girl may dream of a headmistress being killed by the girl's best friend, representing the girl's unconscious desire relating to her mother – a double displacement. The dream thus gives effect to the wish, but at the same time without incurring the disapproval of the preconscious as it would if it said what it meant.

Condensation is another form of dream work. A dream is often very brief, and its interpretation, as Freud says, may be five times as long.[2] A dream of four lines on page 264 of his book takes nearly three pages to interpret. There are many forms of condensation in symbols used in ordinary life. The three stripes on the collar of our sailors are a reminder of the three great naval

1. *The Interpretation of Dreams*, p. 260.
2. *Ibid.* p. 261.

victories of Nelson. (The American sailors also have them!) But it would take a long time to give a full account of those victories they represent: they are condensed into three lines. In the same way the little house between the two stately palaces of the dream on p. 19 represents not only the sexual approach but a place in Prague, and the girl whom the dreamer would like to have seduced came from Prague. The dream of Largs (p. 10) is another instance, the one word representing serious problems not only in the life of the dreamer but in that of her sister. Dreams, like symptoms, are 'over-determined', so that one symptom or symbol may represent a multiplicity of ideas and emotions. In dream interpretation we require to discover all the experiences which the symbol represents.[1]

MEANS OF REPRESENTATION IN THE DREAM

The dream, being visual, representational, and non-logical, cannot express such ideas as 'if', 'because', 'as though', 'although', 'either – or'.[2] How then are these expressed? Logical connexions between ideas are expressed in the form of simultaneousness in the dream – just as a painter might represent 'the School of Athens' by painting in one picture all the poets and philosophers, although in fact they lived at different periods. 'Cause and effect' may be represented by one dream (the cause) followed by another dream (the effect). 'That part of the dream which is more completely worked out always corresponds to the conclusion.' Cause and effect may also be represented by something changing before our eyes in the dream. 'Either – or' is represented in the dream as 'both'. Freud here takes the illustration of his own dream of Irma (p. 152). He excuses himself in the dream for his failure to cure her on the grounds that either it is her own fault, *or* that it is an organic condition for which he is not responsible, but in the dream *both* are asserted. Similarity, agreement, or community are expressed in the dream by unity. Symbols are means whereby we express and convey ideas, feelings, and

1. A symptom of *suffocation* in a patient I am treating at the present moment represents (*a*) an actual suffocation by a nurse in infancy; (*b*) a sexual orgasm in which breathlessness is a common accompaniment; (*c*) a sense of escaping from life and its responsibilities by means of illness; and (*d*) a form of self-pity and looking for the sympathy of others.

2. *The Interpretation of Dreams*, p. 290.

emotions. Words are the most common form of symbol used for this purpose; but in dreams the words have to be turned into visual images. In the dreams there is often a reversal of this process – from the word we should use in ordinary life to the concrete symbol from which it was originally derived – as in the phrase 'You make me sick!' Freud points out the fact that language has at its disposal a great number of words which were originally intended in a picturesque and concrete sense, but are at present used in a faded abstract sense. This makes it very easy for the dream to represent its thoughts. The dream need only restore to those words their original significance or follow the evolution of their meaning a little way back.[1] For instance, we speak of a person being 'in a hole' or in a 'tight place', in which a concrete occurrence is made to do service for an abstract idea. In the dream the order is reversed and the dreamer is represented as literally being put in a tight corner or down a hole in the street. Another instance given by Freud is that if a person dreams of people around being of very big size, it means that it refers to a dreamer's childhood when all around were big. We might suggest that it might also symbolize that other people were superior to the dreamer – as in the phrase 'he made me feel small' – a case of condensation. It is easy to see the part puns play in dreams. A man dreams that his uncle gave him a kiss in an automobile. The dreamer immediately gave the interpretation – 'it means auto-erotism' – and raised the whole question of his masturbation and self-love. Here we have an instance of displacement, of condensation, and of the use of symbol.

For further examples of 'dream work', the reader is referred to Freud's *Interpretation of Dreams*, Chapter vi.

These mechanisms of dreams, discovered by Freud, are largely accepted not only by his followers, but by other scientific schools of thought, as an essential feature of dream interpretation.

OBSERVATIONS ON FREUD'S DOCTRINES

By means of these processes of dream distortion, symbolism, and the rest, Freud seeks to maintain his thesis that, however contrary to appearances, every dream is a wish-fulfilment. He says that 'the fulfilment of a wish is its [the dream's] only purpose'. 'Even

1. *The Interpretation of Dreams*, p. 325.

dreams with a painful content are to be analysed as the fulfilment of wishes.' He concludes, therefore, that 'the dream is the [disguised] fulfilment of a [suppressed, repressed] wish'.[1]

We must admit, however, that in the illustrations he gives to prove his point he is sometimes hard put to it to extract such an interpretation from the dream. His ingenuity in proving his point goes to such lengths that when a patient produced a dream that contradicted his theory, Freud said that the wish-fulfilment in the dream was 'his wish that I should be wrong!'[2] Again, when a lawyer friend dreamt that he had lost all his (legal) suits – certainly not a wish – Freud interprets this as that the lawyer wished that Freud himself should be disgraced.[3] We do not say that Freud is necessarily wrong in these particular interpretations – he knew the patients and we do not. We only say that if we are permitted to resort to arguments of that sort we can prove almost any thesis to be true. By the same token Adler proves that everything is the urge to power. According to him even sexuality is a desire to overmaster the other person; and if it is objected that woman in her sexuality desires to be overmastered, Adler maintains that by submitting she gains power over her mate and gets her own way. So might one maintain with equal justification that the dominant motive in life is that of curiosity, that it is curiosity to have new experiences which makes people indulge in sex, or for that matter seek the exercise of power. Once we accept such a theory all our associations may be 'adapted to the representation of such facts' – to use Freud's phrase already quoted.

As a test case let us take a dream quoted by Freud to establish his point.[4] 'I dream that I am walking in front of my house with a lady on my arm. Here a closed wagon is waiting, a gentleman steps up to me, gives his authority as an agent of the police, and demands that I should follow him. I only ask for time in which to arrange my affairs.'

On being pressed as to the nature of the crime for which he was arrested in the dream the patient admitted it was 'infanticide'. The circumstances were that he had spent the night with this married woman and feared that the contraceptive safeguards had not been adequate. Freud represents the dream as being a wish not to have a child (the dream exaggerated this as infanti-

1. *The Interpretation of Dreams*, p. 136.
2. *Ibid.* pp. 128 and 134.   3. *Ibid.* p. 128.   4. *Ibid.* p. 131.

cide). No doubt there was that wish, but of that the patient was all too conscious: it did not require the dream to tell him. The focal point – indeed the whole substance of the dream – was his arrest by the police agent. This represented his guilt, which he was attempting to evade; but the dream will not allow him to do so. Indeed in the dream he tacitly admits his guilt, for he only asks for time to arrange his affairs. Why ignore this element of guilt in the dream as if it were only a casual intervention on the part of the pre-conscious and say as Freud does that the fulfilment of a wish is the *only* purpose of such a dream? The dream obviously deals not merely with the wish but with the fear or wrongness of that wish; it warns him of the consequences resulting from that wish, and of the punishment with which his conscience threatens him. The man was quite aware of his wish; it was in full consciousness, and not an unconscious wish, which alone, according to Freud, expresses itself in a dream. What the dreamer was unwilling to face was his guilt, which indeed was evidenced by his reluctance at first to admit the facts of the case to Freud. What the dream tells him is that if he carries on like that by having these illicit affairs he will have to face the consequences, which the dream dramatizes by having him arrested. The wish was there, but, far from being a fulfilment of his wish, the dream obviously condemns his wish and his lack of responsibility in no uncertain terms.

Freud of course recognized this conflict in the dreams producing a state of anxiety, but he regards this condemnation as simply the work of the pre-conscious frustrating the full expression of the wish and therefore not of the essence of the dream. As we see it, the guilt is the essential feature of the dream, which the patient was evading and which therefore it was the function of the dream to convey. It is the dreamer's conscience, not his wish, which is repressed and which is the dominating factor in the dream. Freud later came to recognize this fact, but only after others had called attention to it (see p. 53). The dream, then, is obviously concerned not only with a wish but with the fear or wrongness of that wish: it not only presents us with a desire but with a problem arising from that desire. Therefore while Freud has done an incalculable service in calling attention to the latent and hitherto unrecognized wish element that lies beneath the fear, he is not justified in extracting this one element of the dream as though that is the only important aspect of the dream, and saying as he does that the fulfilment of a wish is the dream's only purpose, whilst ignoring as an essential feature of such dreams what is their dominant function, namely the stressing of the sense of guilt. We must take the dream as a whole, and the inhibitions, fears, and guilt of a dream are just as important factors as the wish element. In other words, the subject-matter of a dream is not only a wish but *the problem created by that wish*: that is what the dream is about, that is the problem presented by the dream, which the dream tries to solve.

This leads to another variant from Freud's views, namely that the unpleasantness of dreams is not merely because the pre-conscious objects to them, but because they deal with unpleasant problems like guilt which we are trying to evade but which the dream thrusts upon us for attention and solution – a theme to which we shall return in presenting the Biological Theory in a later chapter.

Again, what appears in the dream to be a simple wish according to Freud's view may in fact be merely a statement of what is taking place in the psyche. A man dreams that he is having a bath naked in the presence of a washerwoman, and that he does not mind and that she takes no notice. This might be taken to represent an exhibitionist tendency in relation to his mother: and so in part it is. But the crucial point of the dream (in contrast with what would happen in an actual occurrence of that nature) is that neither of them takes any notice. If it had been a mere wish, no doubt the washerwoman would have taken notice! This patient always had a feeling of shame about sex since as a child he was reproved for exhibitionism, and this shame had in fact appeared in earlier dreams. This dream is counteracting this false feeling of shame regarding sexuality given him as a child, and it does so by reproducing the original situation with his mother, but taking up a more natural attitude towards sex – nobody bothers about such things – there is no need to be ashamed. This dream, unlike the last, does not impress his guilt upon the dreamer, but on the contrary is ridding him of his false sense of guilt and shame.

But even if we agree with Freud that the sole function of a dream is to give expression to a wish, we must ascribe to the term 'wish' a wider significance than that merely of wishful thinking or hallucinatory gratification, which by the release of tension permits us to sleep. For what is the biological function of an ordinary wish? It acts as an incentive to action. It keeps before our minds something we want, and therefore something we wish to aim at, to work for, to achieve. When we are dispirited in some endeavour, the very fact that we keep the idea in consciousness as a wish acts as a spur. That was true of Joseph's dream of the ambition to surpass his brothers, with their sheaves bowing down to his sheaf; it acted as a spur to action and led to actual achievement. Such wishes, it is true, whether in the day or night, may remain nothing but wishful thinking and lead to no action but only to disappointment and disillusion, in which case they are detrimental to endeavour: we have a beautiful dream and wake up distressed to find that it was only a dream. But in many other cases the dream has incentive value in keeping alive a desire which in the day we may have abandoned as unattainable: it urges us to try again.

Thus men who have abandoned projects have dreams of success which have urged them on to try again. If the dream of Bruce and the spider is a fiction, it is nevertheless psychologically true. Many 'impossible' ambitions have thus been achieved. Even the dreaming of sumptuous meals by the Arctic explorers may encourage them to go on by reminding them, when they are likely to be discouraged, of what is in store for them, and save them from defeat when the hardships of the day would lead them to despondency and despair. When the downtrodden underling has to yield to circumstance and to take things lying down, he dreams of himself standing up for his rights and being victorious. He did not realize that he had it in him until he dreamed of himself actually living that role. The very fact of having that dream boosts his self-esteem: he squares his shoulders, is encouraged to have another shot at it, and he comes out successful. Thus a dream is not merely a wish, it is an encouragement and inspiration. Because it shows a man what he can be, it enables him to become so. The dream is not merely wishful thinking, it is creative and purposive; it does not merely allow us to sleep in the night, it encourages us to action in the day. Thus dreams have a biological and not a mere hallucinatory function to perform.

Freud's views would probably have gained more acceptance if the word wish had been given wider significance and not been regarded merely as a hallucinatory gratification – for all dreams, according to Freud, are hallucinatory.[1] It would have been better if a concept like 'urge' or 'potentialities' had been used in the translation, or even the psychological term 'conative tendencies', or 'goal-directed activities', for these terms have a deeper biological significance than the term wish. There are a number of biological urges in human life, one of the most important of which is the sexual, but others being the aggressive, the maternal, hunger, the urge to seek food, the urge to curiosity about anything new or strange, and the urge to power, all of which and many others are necessary for the survival of the individual and the race, and therefore demand expression. All of these are subject to frustration, all of them may give rise to conflict, and all of them, therefore, can create serious problems which can emerge in our dreams, and which, as we shall see, it is the function of dreams to work out.

Regarded in this way such urges are not only what we wish but what we need in life; the wish is, in fact, the expression of a need. That is the turn that Jung has given to the idea – that dreams are not merely the exercise of a repressed wish but the expression of strong potentialities in our life, expressive of what we need for life, and we do well if we heed such dreams. This gives to dreams a positive value, and leads to the compensatory view of dreams as maintained by Jung, to which we shall return later.

We may summarize, then, by saying that Freud has opened out to us the realm of dreams in a way that no one had previously done, and encouraged the study of dreams by demonstrating the rationality which lies behind their bizarre manifestations, by pointing out the wish element which lies hidden behind even the most horrible and distressing dreams, by stressing the dynamic nature of mental processes in general and the purposive nature of dreams in particular, and by describing the mechanisms of dream formation or 'dream work'. But his wish-fulfilment theory would be much more acceptable if in the first place he had not kept so rigidly to the theory that all dreams relate only to sexual desires; if he had recognized the fact that the subject-matter of dreams is not merely a wish, but the problem created by that wish, and if he had given to the term wish a wider significance as a biological urge or potentiality, and so found in dreams a more

1. *Introductory Lectures*, p. 110.

positive and creative contribution to our lives. It is hardly to be expected that Freud, even with his monumental discoveries regarding dreams, should be able to comprehend the whole truth at once; nor does he claim to do so. What he has given is enough to establish the study of dreams on a scientific basis, and it is on that foundation that others have built.

LATER FREUDIAN THEORIES

That these criticisms are justified is proved by the fact that Freud, in his later edition of *The Interpretation of Dreams,* himself considerably modified his earlier views, but we deal with these modifications separately because they are largely an adoption of views in the meantime already anticipated by Adler and Jung as against Freud's view. It indicates Freud's open-mindedness that he was thus able to change his views, but one could wish that he had made some acknowledgement of his indebtedness to these authors, for he could not have been unaware of their views. In fact in *The Interpretation of Dreams* he never mentions Jung except to quote the latter when he confirms Freud's own views!

In the first place, he later recognized the *aggressive* element, as well as the sexual, as a primary factor in the development of the neurosis; and presumably he would also admit this as a factor in dreams and as an object of wish-fulfilment. Indeed it has already struck us as curious that in *The Interpretation of Dreams* (p. 109) he stated that *day* dreams were concerned with a desire for power as well as being of an erotic nature, whereas the dreams of the night were essentially sexual. 'Day-dreams', he writes, 'are literally wish-fulfilments of ambitions or erotic wishes.' His acceptance of the assertive element, of course, brings his theory of neuroses and presumably his interpretation of dreams nearer to that of Adler, who emphasized the urge to power or masculine urge as the main motive in the psychoneuroses. This modification of Freud's previous views is a concession to Adlerian psychology.

There are other modifications which Freud made of his earlier views, as described above, which require to be mentioned. He was well aware that many dreams, indeed the majority, are painful, and this, we have seen, he ascribed to the work of the preconscious, which refuses to allow the forbidden wishes to enter into consciousness, and therefore gives rise to the painful conflicts. The dream, whose function is the gratifying of the sexual wish, fails in its purpose because of the opposition of the pre-conscious, which censors the forbidden wish: what is wished by one system (the unconscious) is rejected and repressed by another (the pre-conscious). This conflict, which is found in dreams, means that instead of producing pleasure the dream produces pain, with the result that the patient wakes up in anxiety.[1]

Other of these painful dreams Freud previously explained as masochistic, namely concerned with the sexual pleasure in suffering pain – just as some people have pleasure in being beaten. He now completely reverses the order and accepts certain dreams as 'punishment-dreams'. He holds that the wish in these dreams is no longer the wish of the unconscious sexual desire, but a wish of the 'ego' to punish one for having that desire. 'I see no objection to regarding such dreams as punishment-dreams, as distinct from wish-fulfilment dreams.'[2] 'Thus it would be the essential characteristic of punishment-dreams that in them it is not the unconscious wish from the repressed material (the unconscious) that is responsible for dream formation but the punitive work reacting against it, a wish pertaining to the ego even though it is unconscious (*i.e.* pre-conscious).'[3] He adds in a footnote, 'Ever since psychoanalysis has dissected the personality into an ego and a super-ego [in his Group Psychology] it has been easy to recognize in these punishment-dreams wish-fulfilments of the super-ego' – which is the moral part of the personality. He now says in effect that what is 'fulfilled' in the dream is not the sexual wish but a moral wish for punishment for the forbidden wish. This he describes as 'something that is new in the theory of dreams'.[4] It is not merely a new idea; it is completely revolutionary as far as Freud's interpretation of dreams is concerned. For no longer is it the function of these dreams to express an unconscious sexual wish but to condemn such a wish; they have a moral function to perform!

But in fact it was not 'something new in the theory of dreams', for Jung had long before maintained, as against Freud's view that the unconscious could only wish, that there could be a *conflict in the unconscious*, such as that produced by guilt. Indeed I was

1. *Introductory Lectures* (1932 ed.), p. 534.
2. *Ibid.* p. 440.
3. *Ibid.* (1934 ed.), p. 514.
4. *Ibid.* (1934 ed.), p. 440.

personally informed by Jung that this was one of the main reasons for the disagreement between Jung and Freud which led Jung to break away from Freud. A person can repress his conscience as he can his sexual desires, and the function of dreams, as Jung maintained all along, and as Freud now maintains, is often to bring home his guilt to the dreamer. Indeed Jung specifically mentions a dream (given on p. 53), to make clear this distinction and to draw the contrast between Freud's wish theory and his own. Jung's concept of one of the functions of the dream being to bring home to a man his sense of guilt is well illustrated in the 'infanticide' dream (p. 26), which to Freud was purely wish-fulfilment, but the purpose of which, to any unbiased reader, and presumably to Freud later on, was obviously to bring home to the dreamer the sense of his wrongdoing.

But even so Freud seems loath to abandon his earlier accepted theory, for he says,[1] 'One must remember that there are masochistic tendencies in mental life to which such an introversion might be attributed.' Freud here regards the punishment as at any rate sometimes masochistic. In fact there is no reason why such a dream should not contain both motives at once, namely as expressing both a masochistic wish and its condemnation in guilt. This would make it a compromise satisfying both the moral conscience and the forbidden sexual desire. A parallel in a neurosis was the case of a patient of mine who used to punish himself for his masturbation by having himself beaten; but this beating also aroused his sex feelings.

A further difficulty remains: if we are now to use the term 'wish-fulfilment' in this wider sense, as referring to the moral forces of the mind as well as to the natural impulses and desires, there is scarcely any meaning in regarding it as a characteristic of dreams, since it is true of all mental processes. For psychologists have long maintained that in every thought, even those regarded as purely intellectual exercises like that of working out a mathematical problem, there is the conative or striving aspect as well as the cognitive and the affective. In that sense not only dreams but *every thought process is a wish-fulfilment*, and to say that dreams are wish-fulfilments is only to ascribe to dreams what is characteristic of all thought processes without exception. As long as Freud set pleasurable wish-fulfilment against the realistic

1. *Introductory Lectures* (1934 ed.), p. 440.

and moral forces which inhibited these wishes, there seems some reason in describing dreams in these terms, but if the term is to be used also of the forces which condemn the wish, the theory of wish-fulfilment in dreams loses its significance, for it only tells us that dreams have a dynamic quality which is shared by all mental processes, conscious and unconscious. We need not therefore labour the fact that dreams are wish-fulfilments as if that were some distinctive feature of dreams, for that is true of every thought we possess. The distinctive feature of dreams, as we shall later declare, is that they present us with problems arising from the conflict between two opposing wishes, and particularly problems which the conscious waking mind has failed to solve.

These changes in the theories of Freud have been reflected in the changing views of his followers. It is interesting to note that after nearly half a century of somewhat blind acceptance of Freud's views his followers are now tending to diverge from them and indeed are splitting up into various schools. We cannot, of course, follow all of these, but let us take as typical of such deviations the views of Dr F. Alexander, a Freudian psychoanalyst and the Director of the Institute of Psychoanalysis in Chicago, who appears to have been greatly influenced by Jung's teaching, though as far as we know, without admitting it. In his excellent book *Fundamentals of Psychoanalysis* he emphasizes the inhibiting factors in the dream as well as the wish element as an *integral part* of the dream, and comes near to saying that the guilt motive may be an essential feature of the dream. 'Dreams can be best understood as the result of two opposite forces: the wish to express a desire or relieve tension and the opposite tendency to reject the desire.'

Alexander goes further: he says, 'Most dreams with an unpleasant conflict are dreams in which the motivating force is not repressed wish, but a guilty conscience.' Note the emphasis on the term 'the motivating force' – in other words, the main *function* of the dream is to give expression to the guilty conscience. This might almost have been taken directly from Jung.

We welcome such modifications of Freud's views as giving a much more realistic picture of the meaning and function of dreams, but we find others of Alexander's arguments difficult to follow; for instance, his well-meaning attempt to bring his views regarding the importance of the conscience and guilt in dreams into line with the Freudian view that the function of the dream is to allow the patient to sleep. This attempt leads him into queer paths. He does this by ingeniously, but somewhat ingenuously, suggesting that a bad conscience implies punishment; that the expected fear is of castration; and that the dream, by suggesting some lesser form of punishment, allows the patient to sleep! This neat argument is unfortunately contrary to the facts, for the punishment meted out in dreams is often infinitely worse than the reality. For instance, the boy who caught pneumonia owing to disobedience had in his delirium suffered all the tortures of the damned in hell. The man who ran the risk of an illegitimate child (p. 26) is, in his dream, arrested for infanticide – surely a more serious crime! This exaggeration is quite a common experience in dreams, the significance of which we shall consider later (p. 72), but it certainly does not let the dreamer off more lightly, nor allow him to sleep.

Alexander again maintains, like Freud, that the object of the dream is to allow us to sleep, but advances a very different reason from that of Freud. Freud said that a dream enables one to sleep by gratifying the sexual wish and so relieving tension. Alexander says, 'A dream of punishment is an attempt to relieve this tension caused by a guilty conscience because conscience can be satisfied only by suffering.'[1] The wish which the dream has gratified is obviously not a sexual wish even in distorted form, but a wish to suffer, and this not a masochistic wish, but a wish to satisfy the conscience by paying the penalty for that wish. The dream sets out to gratify, not a sexual wish, but its opposite, namely the wish to pay the penalty for having that wish and to propitiate for it.

In discussing the question of why we wake up with dreams, Alexander states that 'The presence of the repressed desires may have been so strong that the inhibiting forces were not able to resist and organize it' and that therefore 'in these cases the dream does not fulfil its purpose successfully.' That can only mean that the 'purpose' of the dream is to resist the sexual wishes and keep them repressed, and it is only when it fails and the wishes come up into consciousness that the individual wakes. In other words, instead of the gratification of the wish enabling

1. *Fundamentals of Psychoanalysis*, p. 150.

this patient to sleep, it has the opposite effect and wakes him. This is not only a modification but a complete reversal of previous Freudian views.

These modifications in psychoanalytic theory are obviously tending towards a very different concept of dreams, for it is forcing us to the view that the function of a dream is not primarily the gratification of a wish, except in the very broadest sense, but the presentation and solution of an unsolved problem. Indeed, Freud, in the later edition of *The Interpretation of Dreams* makes a passing reference to the 'problem-solving' function of dreams. This, as we see it, is the fundamental function of dreams, as described in the chapter on the Biological Theory.

(2) ADLER'S THEORY: THE URGE TO POWER

The theory of Adler is also a 'wish-fulfilment' theory, although he does not call it so. But in this case it is a wish for power and superiority, and this as a compensation for a feeling of inferiority. Adler was the originator of the 'inferiority complex', which has had such a vogue. It is the only complex we know which people boast of possessing, and that for obvious reasons; for to say that you have an 'inferiority complex' is to pay yourself a back-handed compliment, for it means that you think much less of yourself than you really are! His original theory was that we start off with an 'organ inferiority', some physical defect like a deformity, red hair, left-handedness, or just general weakness, and to compensate for that we aim at fictitious goals. Thus the Kaiser had a paralysed arm and to make up for it developed a power psychology to conquer the world; similarly, small men are often very important. But in their attempt to compensate, such people often *over*-compensate; they aim too high, fail to attain their goal, and finally have a breakdown. The failure to attain this fictitious goal, according to Adler, produces the nervous breakdown, paralysis, or indigestion, and these symptoms in turn serve as an excuse for not attaining the goal – 'If only I had not this headache, what could I not have done!'

The emphasis with Adler is therefore the masculine urge, the urge to power. Later he maintained that the basic cause of the neurosis was the lack of cooperation as in the 'pampered child'. But he did not abandon his former views and informed me personally that he harmonized them by saying that it was the child with the 'organ inferiority' who refused to cooperate with others.

This urge to power pervades everything in Adler's psychology, which, as distinct from Psychoanalysis (Freud) and Analytical Psychology (Jung), is termed 'Individual Psychology', to give expression to this urge in the individual. Such desires for power are of course a very common form of day-dream; we day-dream of being head of the firm, being very rich, of romantic love affairs, of being honoured in our profession. Such day-dreams of power are of course commonest in children who are helpless to get the things they would like, or to achieve the things they want, but in their desire to 'watch me do this', or 'see how far I can jump', the power psychology is obvious. Later on, when an opportunity is given by a proper education for achievement, though on a more modest scale, the extravagant day-dreams pass. But where an organic inferiority – weakness of the body, lameness, or any handicap – exaggerates the fantasies and craving for power, the actual achievement is less possible, and may end in a break-down. It is this problem which presents itself in dreams.

As with Freud all dreams are concerned with sexual wishes, so with Adler all dreams are concerned with this problem of inferiority-superiority, and the contents of the dreams are in general symbols relating to this problem. Thus special relations of 'up' or 'down' represent situations of superiority or inferiority to another person, whereas 'right' and 'left' symbolize personal relations of advantage or disadvantage. Victory may be represented by a flight in the sky, failure by a fall into the abyss. Even if one has a sexual dream, this expresses symbolically a desire to have power over another. This applies not only to the man, but also the woman; for in her desire to be sexually overmastered by the male she is forcing him to give in to her desires and gets power over him in that way. She stoops, but she stoops to conquer. As Adler has much less belief in the concept of the unconscious than Freud or Jung, the dreams of the night are in the main only slightly more disguised forms of the typical day-dream with the obvious motive of wish-fulfilment.[1]

1. Criticisms of these views of Adler as an explanation of the neuroses are contained in my *Psychology and Mental Health*, pp. 12 and 124.

CHAPTER 3

DREAMS AS ARCHETYPAL

JUNG'S THEORY OF THE UNCONSCIOUS

JUNG'S interpretation of dreams differs widely from that of Freud, this difference depending largely on a different conception of the unconscious. If dreams are the language of the unconscious, it depends upon the nature of that unconscious what the significance of dreams may be. Freud's 'unconscious' consists only of repressed material, that is to say of material once consciously experienced by the individual and then repressed. Moreover, this material which has been repressed consists entirely of infantile sexual wishes. It follows from this that the unconscious can only wish, that it is amoral, and that therefore there can be no conflict in the unconscious. These sexual wishes, however, being incompatible with the self and repressed, can emerge in dreams only in the distorted form of symbols in order to evade the censorship of the pre-conscious, which is the go-between of the unconscious and the conscious. The function of dreams, according to Freud, is to relieve tension by giving hallucinatory gratification to these repressed and forbidden wishes, and so to allow the person to sleep.

Jung's theory of the unconscious differs from that of Freud in most of these respects, and as a result so also does his conception of dreams. Jung agrees with Freud that individual experiences of the past may be repressed, and that therefore part of the unconscious consists of these repressed personal experiences. Jung calls this the 'Personal Unconscious'. But he maintains that a study of dreams as well as of the delusions and bizarre ideas of the insane clearly proves that there is material in them that could never have come from personal experiences but which must be derived from much more archaic sources and levels of the mind, the origin of which can only be found in our primitive ancestry and in the experiences of the race. 'The unconscious', says Jung, 'displays certain contents that are utterly different from those of consciousness.'[1]

He thus distinguishes between the personal unconscious,

1. *The Integration of the Personality*, p. 6.

which consists, as with Freud, of what has been conscious and then has been repressed, and the racial or collective unconscious, which consists of the residue of all our ancestral experiences, which manifest themselves in primordial images. Jung says, 'I have chosen this term "Collective" because this part of the unconscious is not individual but universal; in contrast to the personal psyche it has contents and modes of behaviour that are more or less the same everywhere and in all individuals.'[1]

ARCHAIC IMAGES

Evidence suggesting this is found in our bodies, which carry about with them the traces of their archaic ancestry. Our aquatic ancestry, for instance, survives not only in the persistence of gills as the Eustachian tube connecting the mouth with the ear, but also in the fact that in our veins there flows a stream of blood plasma which consists of practically the same constituents, sodium, potassium, and calcium, and in nearly the same proportion, as sea water, which it originally was. As with our bodies, so with our minds – we carry about with us archaic modes of thinking and of behaviour. Have we not all at times been astounded at the bizarre thoughts that occur to our minds, at the shocking desires which sometimes possess us, and at the irrational impulses that sway us against our volition? They come 'out of the blue' of our racial unconscious and are quite alien to our ordinary civilized modes of thinking. It is not surprising that St Augustine thanked God that he was not responsible for his dreams!

This collective unconscious, says Jung, consists of the 'inherited potentialities of human imagination',[2] 'It is the all-controlling deposit of ancestral experiences from untold millions of years, the echo of prehistoric world events to which each century adds an infinitesimal small amount of variation and differentiation.' 'These primordial images are the most ancient, universal, and deep thoughts of mankind.'[3] The collective unconscious therefore 'contains not only every beautiful and great thought and feeling of humanity, but also every deed of shame and devilry of which human beings have ever been capable.'[4] The

1. *The Integration of the Personality*, p. 52.
2. *Analytical Psychology*, p. 410. 3. *Ibid.* p. 411. 4. *Ibid.* p. 414.

collective unconscious constantly affects our habits and behaviour quite unknown to ourselves, for 'it is a determining constituent of all experiences.'

Jung does not maintain that these ideas and images are themselves handed down as such, but their potentialities are. They are not 'transmissions of representations, but only the possibility of such representations, which', as he rightly adds, 'is a very different thing.'[1]

These potentialities in the collective unconscious have not been repressed in the Freudian sense, since they have never yet been experienced in consciousness. They are a 'flood of thoughts and experiences that have never been in the individual's mind, nor that of his doctor nor in any other normal mind.' 'They are the products of an autonomous independent mental functioning never before known or experienced.'[2]

Being archaic, these potentialities are of quite a different nature from thought processes in the conscious mind: they are subject to different laws. They are quite irrational, illogical, and for the most part amoral, although morality and guilt are also archaic in origin and character. But these potentialities in the collective unconscious are capable of invading conscious life and often determine our fate. They produce dreams, irrational fantasies, peculiar visions; it is they which appear in a form of primitive emotions, grotesque and fabulous ideas.[3] They appear in nightmares in the form of terrifying creatures, now as dragons, now as serpents, now as monsters of the deep or as blood-sucking vampires which disturb our sleep with horrifying dread. It is not surprising that man has come to fear the unconscious, and such fear is a constant constituent of dreams. We dream of having to go into deep caverns of the earth, or stand on a shore to see some monstrous octopus threaten us from the depths of the sea.

Even civilization's discoveries of science have not delivered us from these fears of the unconscious; the philosopher and the scientist are as much prone to irrational fears as the rest of us. For 'the enlightenment which stripped nature and human institutions of the gods overlooked the one god of fear who dwells in the psyche.'[4]

1. *Analytical Psychology*, p. 410.
2. *The Integration of the Personality*, p. 7.
3. *Ibid.* p. 17.
4. *Ibid.* p. 293.

Dreams as Archetypal

Apart from their emergence in dreams, these primordial and entirely irrational forces, surging up into conscious life, are so strong and autonomous that they may become a positive danger to the community as well as to the individual, and that is why they have to be suppressed by the demands of civilization. In the international field they break out in irrational wars which nobody wants, in strife and destruction between nation and nation. In national life they break out in revolutions; in social life, in periods of libertinism and unrestraint, in black-marketing and gangsterism. In industry they give rise to unrest and strikes. Of this we have witness in the present day, for all these disasters emerge from time to time in the world's history as the result of the eruption of these volcanic forces in the collective unconscious, in spite of all the efforts of civilization and reason to keep them in check.[1]

In the individual they produce crime, for the murderer is probably right when he says that 'something came over me' which he could neither understand nor at the moment control, and which gave him a 'black-out'. In the individual once more this conflict may take the form of psychoneurotic disorders in which the depressed forces and instincts surge up to produce anxiety states, hysteria, and obsessions. In extreme cases where the individual is not sufficiently strong constitutionally, these unconscious forces may break through and so unbalance him that he becomes psychotic or insane, completely possessed by bizarre delusions, uncontrollable passions, violent actions, and degraded habits altogether alien to normal behaviour. 'The lunatic is an individual completely overcome by his unconscious.'[2] We gather, however, that Jung regards the neuroses as emerging more from the personal unconscious and its repressed experiences, and the psychoses as emerging from the collective unconscious, for he says, 'The unconscious neurotic conflicts are within reach of consciousness; they are humanly understandable; the unconscious material of the psychoses [insanity] is not understandable'[3] – because they arise from the irrationalities of a collective unconscious. 'A psychosis is a largely involuntary yielding before an irruption from the unconscious that has attained a higher potential than consciousness and so overflows the

1. *The Integration of the Personality*, p. 293.
2. *Ibid.* p. 8. 3. *Ibid.* p. 7.

inhibiting barrier.' Therefore in a psychosis we are completely overwhelmed by our irrational thoughts and emotions, whereas in a neurosis which comes from the personal unconscious we can keep a certain check on them. That may be why a neurosis is so much more accessible to treatment than a psychosis.[1]

But, as we shall see, these forces in the unconscious, powerful and dangerous as they may become, may be turned to good as well as to evil, for the unconscious contains all the potentialities of life, and is the source of energy, the reservoir of all power from which we draw our daily strength. It contains the *material* of our lives, it is the mother from whose fertile womb spring all the forces of life and energy.

THE ARCHETYPES

The *forms* in which these primordial images manifest themselves in conscious life Jung calls 'Archetypal Images' and the potentialities themselves the 'Archetypes'. Archetypes may be described, therefore, as the forms taken by the archaic potentialities in the collective unconscious.

Jung's idea of the archetypes differs considerably from the idea of the instincts as described for instance by McDougall.[2] Both are universally transmitted potentialities, but the instincts are inherited patterns of behaviour like fear and sex, whereas Jung's archetypes consist rather of images, or 'thought-feelings' as he puts it elsewhere, patterns of thought such as we find in myths and in dreams. It is one thing to have an inherited instinct to run away from danger, quite another to have an inherited image of a monster, or at least a potentiality towards such an image. It is true that he says that it is not the ideas or images themselves which are transmitted but their potentialities, but these potentialities are nevertheless of specific ideas and images.

Instances of the archetypes, constantly appearing in myths and in dreams, are the Hero (the Deliverer, the Saviour, the Messiah, the Man of the Moment, the 'Strong Man' that the nations of Europe in their distress are looking for), the Terrible Mother (portrayed as the witch, the avenging goddess, the cruel stepmother, the ogre), and others. These ideas or images, constantly recurring in dreams and in myths, represent

1. *The Integration of the Personality*, p. 90. 2. *Social Psychology*.

deep-seated racial experiences in man.

The idea of sacrifice is archetypal, the need for sacrifice being a basic factor in life throughout nature, the sloughing off the old for the new, the sacrifice of a good for a greater good. In mythology the idea is embodied in the Phoenix, whose body is burned to give rise to a new being. Sacrifice also appears in most religions as a propitiation for guilt, showing how deep-rooted is this idea of guilt in human nature, as well as the need to propitiate.

Another archetype is that of the *shadow* which is the undeveloped aspect of our personality, 'the inferior or less commendable part of a person',[1] which is constantly personalized in dreams. What Jung calls the 'shadow' is reflected in such sayings as, 'He forgot himself', 'He behaved in such an odd way', 'Wait till you know him better', 'The meeting with oneself is the meeting with one's own shadow' – it is an 'other self'. 'The shadow ... is the historically older human being'[2] – it represents the puerile, the childish, immature part of ourselves. When therefore an individual meets with circumstances too difficult for him, he tends to 'regress', to sink back into the puerile, into childish reactions, and this constitutes a neurosis.

Another interesting example of an archetype is that of the *Persona*. Since we have to live in a social world, and yet have undesirable traits in our characters which we are unwilling to show to others, we have to put on a mask to hide our real selves, a mask of bravado to hide our fear, of indifference to hide from others how much we feel, a mask of politeness to disguise our hate or contempt. This Jung calls the *Persona* (= mask). Like the *Dramatis Personae* of a play, we are not really such persons but are pretending to be.

The figure of this *persona* is constantly coming into dreams, for the conflict between the real self and the *persona* is a common problem and a frequent cause of neurosis.

Another archetype which is of great significance is that of the *anima* which often appears in the man's dreams as a female form, and in the woman's dreams as the *animus*, represented as a man. Every human being has both masculine and feminine characteristics, both physically (as with breasts in the male and clitoris or rudimentary penis in the female) and psychologically. Thus some men are smooth-faced, gentle, and effeminate; some

1. *The Integration of the Personality*, p. 20. 2. *Ibid.* p. 91.

women moustached, masculine, and bossy. If one is a male, that is to say a person in whom male characteristics predominate, then the female tends to be suppressed, but this suppressed feminine part of the male, which Jung calls the *anima*, appears in his dreams as a female form. Similarly woman has her rudimentary masculine tendencies suppressed and these appear in dreams as a man: her *animus* is the man in her. Jung says, 'The feelings of a man are, so to speak, a woman's and appear as such in dreams. I designate this figure by the term anima; she is the personification of the inferior functions which relate the man to the collective unconscious as a whole and which presents itself to the man in feminine form. To a woman it appears in masculine form and there I call it "animus".' An *anima*, then, is the feminine aspect of an individual man, 'the woman in a man'.[1] But 'it is not always merely the feminine aspect of the individual man'; it is the 'eternal feminine', far older than that of the individual; it is therefore an archetype.

Animus and *anima* are figures which constantly appear in mythology. They 'are natural archetypes, primordial figures of the unconscious, and have given rise to the mythological gods and goddesses.' 'They live in a world quite different from our own.'[2] 'It is therefore rather a futile undertaking to disinfect Olympus with rational enlightenment' – for men and women will continue to believe in gods in spite of all scepticism and rationalism because they stand for something real in themselves!

In ancient mythology the *anima* is represented by such figures as Helen of Troy and Venus;[3] in medieval times by Mother Church or the Queen of Heaven. In modern times it has been vividly represented in Rider Haggard's 'She', in Faust's 'Gretchen', as representing the fallen woman, and in the Virgin Mary representing the immaculate woman. The *anima* is also represented in Tennyson's 'Lady of the Lake', who 'knows a subtler magic'; 'dwells down in a deep; calm whatever storms shall shake the world' – 'a mist of incense curl'd about her, and her face well-nigh was hidden.' All these figures obviously have reference beyond the personal and individual, for they stand for the eternal feminine, the ideal woman, the 'woman of my dreams'.

While the *anima* appears in a man's dream as a woman, this

1. *The Integration of the Personality*, p. 73.
2. *Ibid.* pp. 23 and 25. 3. *Ibid.* p. 78.

does not mean that every woman in a man's dream necessarily represents the *anima*, nor does Jung maintain this. But if a man is unconscious of his *anima*, it is frequently projected upon a woman, perhaps a real woman of his acquaintance, who is chosen as characterizing his feeling; or just a woman, who may be either a depraved woman or the feminine imperatrix, according to the nature of the man's *anima*.[1] 'The woman of our dreams' often takes on an idealistic form, and many boys and girls fall in love with an ideal which is the projection of their *anima*. They are the kind who never marry because they 'never meet the right girl' or 'right man', that is, one who comes up to their fantastic dream, or the narcissistic fantasies of their own worth. Or it may be that they endow a quite ordinary person with these idealistic qualities, so that the boy sees in the gawky girl a perfect Venus, and the girl sees in the pimply boy a perfect Adonis – only to be disillusioned later.

As long as these functions are suppressed the personality is incompletely fulfilled; for the female characteristics or *anima* in a man are necessary to the fulfilment of his personality, and the masculine element or *animus* in a woman is equally necessary for her complete development, whether to manage her household or to do some other job in life effectively. When, therefore, they appear in dreams, they are complementary to our ordinary waking selves, and represent what is necessary for the fulfilment of our lives. According to Freud, to dream of a woman refers to one's sexual desires; according to Jung, it may be sexual, but equally it may refer to an unfulfilled and undeveloped part of our personality, necessary to the fulfilment and completeness of our lives, and we should take heed to such dreams.

Another instance of the archetype is that of the 'wise old man', the teacher, the sage who is the embodiment of the wisdom of the ages. In our day we tend to associate the idea with the Lama of Tibet (perhaps because of the remoteness and inaccessibility of that region); or it may be with the Pope, whose dictates are accepted as authoritative for millions of people; or the Archbishop with his 'pronouncements'. When bewildered with the problems of life, we look for some such authority for a teacher and guide. Indeed we may venture that Jung himself embodies this archetype to some of his followers; and very worthily he

1. *The Integration of the Personality*, p. 23.

fulfils that role! But there is an 'ancient wisdom' in us all which, in the ordinary affairs of life, goes beyond what we call 'common sense', and acts as a guide to us in the more fateful problems of our lives, and which in our dreams may appear as the old wise man, the patriarch, the father. We do well to heed what he says, for intuition, so often represented as a wise old man, is often a better guide than reason.

These archetypes, relating as they do to deep-seated ideas and feelings in the collective unconscious, cannot adequately be expressed in the language of reason and therefore take form in fairy stories, in myths of the race, and in dreams, and also find expression in music and the primitive as well as in modern art.

There would be no function for the artist and musician to perform if all that we experience could be expressed in words, for art and music can give expression to feelings which cannot be otherwise expressed, nor otherwise appreciated. So these archetypal images can sometimes be best expressed in symbolic art forms, which are pictorial representations of what is going on in the personality and especially in the unconscious. This is particularly obvious in the automatic drawings of the insane, from which a study of the archetypes may be made, for in them have appeared strange ideas and imaginings which never were on land or sea; also in the automatic drawings of children through which we may interpret the child's difficulties and problems. Such drawings are constantly used by the physician, for by them he can diagnose the nature of the patient's disorder.

Many archaic modes of thought are to be found in religious practice, which can be expressed only in symbol and justify the use of ritual in religious worship. The idea of baptism represents new birth, and the sacred meal which in primitive life takes the form of eating the totem animal of the tribe, and thus 'eating the God',[1] persists in the Christian Sacrament, which some still take to mean the eating of the actual body of Christ, or God; whereas others take it more symbolically as representing partaking of the life of God, the source of all life and goodness, or the acceptance of Christ's sacrifice. It is impossible to express these experiences adequately in words, because they are mystical and matters of feeling rather than of thought, and they must therefore be expressed symbolically. To attempt to tie them down to ordinary

1. See Robertson Smith's *Religion of the Semites.*

and logical thinking would be to destroy their meaning. In any case, in many people the symbol or ritual is capable of arousing deeper emotions and feelings than the expression of the corresponding idea in words, for by the expression of such ideas we gain in clarity but lose in depth of feeling and emotion.

Archetypal ideas emerge also in fairy stories: they depict men fighting against giants, like Jack the Giant Killer, St George and the Dragon, Beowulf, and others, all of which represent man's struggle against the primitive force in his unconscious. There are stories such as Joseph in the Old Testament, despised of his brothers, reaching to great fame; and the story of the Princess shut up in the Tower who has all she needs except that which she needs most of all, namely love, but she is ultimately rescued by the hero; and the story of Cinderella, which originated in China and has travelled all over the world, depicting the theme of the ignored and neglected person coming at last into her own. All these stories relate to common feelings, problems, experiences, and cravings of the human soul. Therefore human beings love to hear these stories over and over again, just as children love to hear the stories such as Little Black Sambo over and over again until they can repeat them word for word, because they relate to some experience or problem in themselves which they but dimly appreciate. Dreams also, as we shall see, repeat the same experience, or the same theme, over and over again because they represent and help to clarify some deep-seated problem in us.

These archetypes also appear in myths, which are the attempts to give expression to collective experience, attempts to solve racial problems, and these also are repeated over and over again, and from generation to generation. A well-known illustration is that of Pygmalion and Galatea, which gives expression to latent potentialities in ourselves and of petrified feelings which come to life and warmth within us. The myth of Prometheus likewise is expressive, among other things, of the rebellion of youth against authority and of its consequences, as is the story of Lucifer's rebellion against God.

What myths are to the race, dreams are to the individual, for in dreams, as in myths, there also appear those primitive emotions and feelings in the form of giants, heroes, dragons, serpents, and blood-sucking vampires; representations of guilt, retribution, and fate; of lust and of power, of monsters of the deep (the unconscious) and of unknown but overwhelming beings which fill our nights with nightmarish dreams and make us fear to sleep, but which, rightly used, can be fruitfully integrated into our personality.

So close is this parallel between myths and dreams that Jung often has resort to myths to help in the interpretation of dreams. This is a method complementary to that of free association, for it may be maintained that even free association cannot plumb the depths of the unconscious mind. When, therefore, the symbolism of a dream is in doubt, corresponding myths suggest a true interpretation. Indeed Jung maintains that the symbols used so readily in Eastern religions originally came from dreams and visions, and were not the invention of some church Father.[1] The symbolism of the cave, referred to later (p. 166), often appears in mythology and also in dreams as representing the unconscious with its dark secret recesses. Similarly, to dream of a return to the womb may represent a need to get back to a condition of safety and security; but it may mean a return to the womb of the unconscious out of which springs all life. There is also the symbolism of the snake, which is often depicted as coming out of the cave, and which Freud would interpret as a phallic symbol, the cave as representing the uterus. But in mythology the snake is often the symbol of healing. The serpent which was lifted by Moses in the wilderness so that whoever looked upon it should be cured of their diseases was a symbol of healing and was adopted by Aesculapius in the Temple of Healing. It now appears in the badge of the Royal Army Medical Corps. The snake is also the symbol of rebirth, because of the sloughing of its skin. In mythology it is the symbol of wisdom. But before proceeding further let us give a simple dream which illustrates various points in Jung's method of interpreting dreams. We shall call it 'the dream of the evil'.

My patient had had her emotional life repressed as a result of her father's unkindly treatment of her, with the result that she became over-intellectualized, and assumed the attitude of the sweet, adoring daughter in order to ingratiate her father. She dreamt that she was blind and had on a very beautiful veil covered with flowers which would have prevented her seeing

1. *The Integration of the Personality*, p. 127.

even if she had not been blind. Then there appeared a Negro, who leant forward and kissed her on the forehead, whereupon her veil disappeared and her sight was restored.

The Negro in the foregoing dream might, on a Freudian interpretation, be the dreamer's father, the kiss representing sexual intercourse; the dropping of the veil, the breaking of the hymen, and the restoration of the sight, the opening of her eyes as having tasted of the fruit of the tree of knowledge. This interpretation is supported by the fact that she had had her sex feelings aroused when her father beat her. It is easy if we start off with a specific theory to make the symbols fit. Jung grants that we may thus interpret such a dream, that is to say, reductively; but he maintains that even in a dream in which the symbols are sexual there is also the *constructive* interpretation. The interpretation might go something like this: the veil with beautiful flowers was obviously her *persona*, the mask she wore, the sweet intellectualized self she had adopted and presented to the outside world. This mask was beautiful and pleasant but prevented her seeing and facing life. The Negro was her *animus*, the crude representation of her primitive and emotional self which she had rejected and repressed. But it requires emotion as well as cold intellect to see the meaning of life. She was therefore blind quite apart from the *persona* she had imposed on herself. It was not until she accepted her crude emotional life, represented by the kiss from the Negro, that the two parts of her personality were brought together, and the false attitude, the veil, disappeared so that she could see clearly and cope with life adequately. The function of the dream is here well represented: it is a compensatory dream in that her intellectualized life, which left her arid (as it did with Faust), needs to be compensated for by her emotional life, so that the two when brought together should make her whole. This patient had previously had terrifying dreams of Nazis and Gestapo men from which she awoke in terror, which partly represented fear of her totalitarian father, but also the fear of her primitive emotions. The present dream is constructive in that it depicts the harmonization between the two parts of her personality. It indicates to the dreamer that if she would accept her primitive self (which, because of her repression of it, had become a source of danger and in fact produced her neurosis) and embrace the primitive in her life, it would mean the restoration of sight so that she could see her way in life clearly and deal with it more effectively. But such a dream is not only constructive but purposive in that it points the way to the restoration of her personality and the cure of her neurosis.

Dreams, therefore, whilst they reveal to us these primitive forces in the unconscious which may be a danger and threat to our personality, also reveal to us, if rightly interpreted, the nature of the potentialities which are available to us for the uses of life.

This conception of these powerful collective forces in the unconscious and of their positive, creative function in life is perhaps Jung's main contribution to psychological thought. Civilization, he maintains, may be over-zealous in its repression of these primitive forces in the unconscious, for the unconscious is the source of all power, and to keep these forces suppressed altogether is not merely to deny to ourselves the power that might be ours, but to invite their hostility so that they become a source of danger. They need to be given expression if they are to be integrated as part of the personality. In this process of integration we are helped by dreams which, by representing what is repressed and suppressed in our life, tell us not only what we want in life but what we *need* in life.

THE COMPENSATORY THEORY

According to Jung, therefore, the function of dreams is that of compensation, namely that the unconscious material which is repressed and which appears in dreams is obviously the opposite of what is in consciousness and therefore compensates for what is lacking in consciousness. 'The standpoint of the unconscious is as a rule complementary and compensatory to consciousness', says Jung, and one of his closest followers in an unpublished paper says, 'The all-important principle of dream analysis is the compensatory function of the unconscious in relation to the conscious.' We see then why 'dreams go by opposites', as the popular saying is. Thus it is the hard man who is most sentimental in his dreams, and the philosopher who is often subject to the irrational dreams, as he may be to the most irrational neurotic fears. The emotions which in his rationalism he has subdued take their revenge by appearing in his symptoms and dreams, demanding by force a place in his life which he refuses voluntarily to give. Regarded in this way, dreams do not merely represent a 'wish' in the Freudian sense, but a 'need'; they express what is repressed in a personality and therefore what we need for its fulfilment.

This principle of compensation operates throughout our dream life. The ascetic dreams of sensuous pleasures; the libertine, as in the infanticide dream (p. 26), of a sense of guilt; the timid man shows fight, and the bully is a coward.

If our experiences in life have led us to repress certain of our functions, and therefore to exaggerate others such as punctiliousness, laziness, dependence, or sensuousness in excess of what is healthy for the psyche, the unconscious in dreams compensates by throwing into consciousness the opposite tendencies and shows us to be easy-going instead of punctilious, hard-working instead of lazy, self-sufficient instead of dependent, or with a sense of humiliation in place of our conceit.

PURPOSE IN DREAMS

This play of opposites may be regarded as a purely mechanical process due to the emergence of repressed tendencies when the conscious control is in abeyance. That may be; indeed we should agree that it is the case. This also applies to many of the functions of our body, such as sight, hearing, and smell; in the process of evolution of these functions they were in the first place mechanically developed as the result of environmental conditions. But what was at first a mechanistic process thereafter became of use to the animal in promoting the welfare of the organism. Thus the eye, which was originally a modification of the skin under the influence of light, later comes to serve the function of sight. We cannot fully understand the meaning of the structure of the eye or of other organs except in relation to the function they serve: in this sense, from being mechanical they became purposive. So it is with these mental processes which emerge in dreams: they may at first be merely the mechanical emergence of potentialities in ourselves which have been repressed. But the material which comes up in dreams, when released, serves the purpose of providing us with what we require for the fulfilment of our lives. On the other hand, if these potentialities are left repressed and unused, they suffer the same fate as the appendix and may become a source of danger to the personality even to the production of neurotic symptoms.

The purposive function in dreams therefore follows from this compensatory nature. Jung says, 'In dream interpretation we ask what conscious attitude does the dream compensate'[1] – and again, 'the dream shows in what direction the unconscious is leading him.'

The compensatory and purposive function of dreams finds its counterpart in dream interpretation. We look to dreams to reveal not only what primitive forces in the unconscious are striving for expression, but also in what respect our conscious attitudes are wrong and our character distorted. In practice, therefore, if a man constantly dreams that he is doing silly and socially stupid things, it probably means that in his ordinary character he is too correct, too conventional, too anxious to maintain his prestige, and too concerned not to make a fool of himself. It may be that he needs to make a fool of himself, to climb down a bit, to be more spontaneous and do silly things for a change, and not to take himself too seriously. The dreams of a man that he is always being held back may imply that he is in too much of a rush and needs to be held back, that he is too anxious to 'get on', and is ruining not only his peace of mind and happiness but his efficiency: he needs to relax; so that when he is too anxious to get on, the unconscious mind holds him back. He must be encouraged to take things more easily. So the emergence of repressed tendencies may be regarded either in terms of 'mechanism', of 'compensation', or of 'purpose'.

Dreams, therefore, according to Jung, are not merely wish-fulfilment; they have a teleological significance, point to an end, they strive towards an attainment of that end; they tell us where the unconscious mind is leading us, they show us the way.

Jung's approach to dreams is therefore entirely different from Freud's. Freud looks at dreams from the point of view of causality, where the dream images came from, what infantile wishes they satisfy, what complexes they reveal. Jung, as he himself says, is not interested in causes but interprets the dream from the view of *finality*, where it is leading us, what it offers for our life. The dream not merely gratifies our wishes, it indicates our needs. Jung admits that the interpretation of dreams accord-

1. *Modern Man in Search of a Soul*, pp. 20, 72, 75.

ing to causality is in accord with the scientific spirit of the time; but the personality is not concerned with causes only but with ends, aims, and purposes, and dreams are as concerned with the direction in which we are going as with the places from which we come. 'Freud', says Jung,[1] 'is seeking the complexes. I am not. That is just the difference. I am looking for what the unconscious is doing with the complexes.' 'I do not want to know the complexes of my patients. That is uninteresting to me. I want to know what the dream has to say about complexes, not what the complexes are. I want to know what a man's unconscious is doing *with* his complexes. I want to know for what a man is preparing himself. That is what I read out of his dreams.'

CONFLICT AND GUILT IN THE UNCONSCIOUS

These considerations lead us to another important feature of Jung's psychology and his interpretation of dreams which sharply distinguishes it from Freud's theory, namely that the unconscious contains not only wishes but moral values and a sense of guilt. According to Freud, the unconscious can only wish, and consists only of sexual wishes. According to Jung there can be a moral conflict in the unconscious itself, and it is this guilt, and not merely the wish, which emerges in dreams, and which it is the purpose of the dream to reveal. Freud recognized such a conflict in dreams, but put it down to the work of the pre-conscious in its effort to keep the undesirable wishes of the unconscious from invading consciousness. Freud, it is true, later came to accept the fact that some dreams are 'punishment dreams' implying guilt, but makes no acknowledgement of the fact that Jung had long before insisted on this fact which at the time Freud stoutly denied. Indeed Jung specifically gives a dream to illustrate this important difference in interpretation.[2] The dream is as follows:

'I am standing in a strange garden, and pluck an apple from a tree. I look about cautiously, to make sure no one sees me.' The patient confessed that he had recently begun a love affair with a housemaid, and on the day before had had a rendezvous with her. This dream obviously contains mythological material which the

1. *Modern Man in Search of a Soul.*
2. *Analytical Psychology*, p. 303, p. 99.

dreamer himself associates with the scene in Paradise – plucking the apple. So far it is wish-fulfilment. But the rest of the dream indicates a bad conscience, a sense of guilt. Freud would admit this, but would ascribe the bad conscience to the work of the pre-conscious. The main purpose, according to him, is to gratify the wish, and it evades the censor by putting it in symbolic form. Jung accepts the wish in the dream – indeed it was obvious that the dreamer wished to possess this girl. But Jung interprets the dream from the point of view of finality; it is the purpose of the dream to emphasize the sense of guilt; to insist that the young man should face up to the conscience he is trying to repress and to the necessity for regarding his erotic conduct boldly from the point of view of morality.

This moral element in the unconscious Jung regards not merely as derived from a repressed sense of guilt in our child-hood, but regards it as something archaic, something archetypal, handed down to us from our forefathers in the racial unconscious.

DISCUSSION OF ARCHETYPES

It is natural in every science that when one is striving to emphasize a new hypothesis or theory there should be some over-emphasis in order to make the point. This to our mind Freud did when he referred all dreams to causes in infant-ile sexuality – a view he later modified; and Jung, in empha-sizing the compensatory and purposive function of dreams, has perhaps too much neglected the causal factors. According to our conception, to be discussed later, every complete and fully-fashioned dream deals both with causes and with ends; for the first part of the dream reveals the problem and indeed often recalls the causes which originally gave rise to it, and then works towards a solution of that problem. That is why we look to the beginning of the dream to discover what is the problem, and to the end of the dream for the suggested solution. This was illus-trated in the dream of the beautiful veil (p. 48). The first part of the dream revealed the neurosis, namely her blindness. Then it gave the cause of the neurosis – that she had assumed this sweet, refined, beautiful mask of a personality; it then led on to the solution of that problem, namely coming to terms with her primi-tive emotions. Freud, to judge by the dreams he interprets in his book, tends to look at the first part of the dream only, a conspicuous example of which is the infanticide dream (p. 26) in which he emphasizes the wish element in the first part of the dream, the wish to be rid of the child, but ignores the end of the dream which deals with the arrest of the dreamer for having that wish. The sting of the dream is in the tail. The dream in point of fact first dealt with his illicit affair (walking with this woman) and ends with bringing home his sense of guilt: it dealt first with the causal factor and then with the purposive factor. The same occurs in Freud's own dream of Irma (Chap. 7) in which he lays all the emphasis on the beginning of the dream, namely the desire to put the blame for his failure in the treatment of Irma on her. Freud maintains that that was the aim or wish of the dream; he fails to interpret rightly the rest of the dream which ruthlessly reminds him of his previous failures and implicitly condemns him for attempting to put the blame on others. Jung, on the other hand, on his own showing, is not concerned with the causal part of the dream but only with the end, where the dream is leading. But if the causes are of no consequence, why does the dream trouble to mention them? In fact the dream often does nothing else but revive the causal factors in experiences of the past, such as being blown up, or terrifying experiences of infancy producing night terrors, and points to no end or solution. We cannot afford to ignore these causal factors as Jung appears to do, and the dream as a rule refuses to allow us to do so; but neither should we, as Freud does, ignore the constructive purposive factor with which the dream often ends.

WHAT OF THE ARCHETYPES?

Whilst we have selected the theme of archetypes for particular consideration as typical of Jung's teaching, it must not be thought that this covers his whole theory of dreams. For instance, there is his insistence on the present-day moral problem as precipitating the neurosis, and which appears in dreams. But the archetypal element in dreams is perhaps the most significant contribution of Jung to the theory of dreams, and possibly the most distinctive feature of his psychopathology. In any case, our object in this and previous chapters is not so much to describe the theories of particular individuals as to call attention to certain aspects of dream interpretation – hence the title of this and previous chapters which represent subjects of study rather than schools of thought.

What, then, are we to think of these archetypes? Are they realities or are they only abstractions? Are they inherited or are they derived from early experiences? Are they collective, existing in some way apart from individuals, or do they merely represent experiences common to large numbers of individuals and therefore taking on a collective aspect? We are justified in considering these problems here, because it is in dreams that the archetypes are most clearly manifested.

It must be remembered that both Jung and Freud have formulated their theories as a result of clinical experience: they describe what they have found on the examination of their patients and in neither case are the theories intended to be systematic, but descriptive. If, on the other hand, we consider these views systematically and logically, there are some difficulties surrounding the idea of the 'archetypes' as racially determined and indeed in the conception of the 'unconscious', to which we must later refer.

Jung has opened the door to a vast *hinterland* of strange forms and bizarre ideas said to be contained in the collective unconscious, evidence of which he finds in dreams, myths, and delusions of the insane, none of which, he maintains, can have come from personal experience and which must therefore be collective and racial. These bizarre ideas are dramatized and personalized in our dreams, since this is the only way in which the primitive mind can give expression to them, but that is no reason why we should *consciously* persist in personalizing them. For clinical purposes it may be justifiable to speak of these aspects of man's personality in this dramatic way as the 'Archetypes', as though the *anima* and the *persona* had independent existences, and this concrete way of speaking is very effective and easily grasped; but it has its dangers, for it gives the impression that they are entities in themselves, instead of being, as they are, only abstractions or personifications of functions of the personality. This is a common habit in psychology. Freud does the same thing when he speaks of 'the censor' as though it were a person sitting aloft and censoring our thoughts and wishes, instead of speaking of 'the function of inhibition'. *We* do the same when we speak of the 'will', as though it were a dictator within us who guides and directs our actions and regulates our conduct, whereas what we call the 'will' is only the activity of the personality as a whole directed towards a common end. Similarly, we speak of conscience, and say that our 'conscience tells us', whereas 'conscience' is only a name we give to the function we possess of self-criticism and self-condemnation: it is no more than we telling ourselves.

There is indeed a strong tendency in human thinking to 'reify', that is to say, to regard mental functions as 'things', against which tendency William McDougall has warned us. If we are to avoid confusion, we must remember that all mental life is a process, a functioning; and we ought not, strictly speaking, to speak of 'the mind' but of mental processes; not of 'the unconscious' but of unconscious processes; not of 'thoughts', or 'emotions', or 'perceptions' as though they were 'things', but of thinking, of feeling, and of perceiving, emphasizing the fact that they are functions. Nobody really believes that there are 'thoughts' apart from a thinking person, or feelings and emotions apart from a feeling person. In brief, in psychology we should always speak in adjectives and verbs and not in nouns, since psychology is the scientific study of mental processes; and we await the attempt by some brave spirit to write a book on psychology on these lines! The archetypes are in the same category as thinking or feeling; they are the name we give to functions and processes we find in the human mind.

The term 'archetype' originated with St Augustine, but the tendency to think that such entities pre-existed before the individual and that they have independent existence is as old as Plato. According to Plato's philosophy, there are 'ideas' or 'forms' of things, created by God, of which all things on earth are patterns or examples. These ideas are conceived as existing in eternity and are the prototypes of all things on earth. When we speak of a 'cat' as such we relate that animal to all other animals of a similar nature, so that by 'cat' we do not necessarily mean this cat or that cat but refer to some general quality of which all cats are partakers – something universal and distinct from the individual, something 'ideal', in fact, and eternal. Nowadays we recognize this as an abstraction, which we might call 'cattiness', except that the term has acquired a meaning expressive of a more undesirable quality! Plato, on the other hand, held that there was in fact an 'ideal' cat created by God of which all other cats on earth are copies or reflections, like the reflections in a mirror. There is only one real cat, in heaven, which is the prototype of all cats that are on earth. Jung in his concept of the archetypes does not, of course, hold to the conception of universal 'forms' or 'ideas' in that crude and criticizable form and differs from it in several respects. For one thing, Plato's original 'forms' were created by God and exist in eternity, whereas Jung's archetypes exist in the racial unconscious; the one comes from above, the other from below. But both contain this idea of universal realities with an independent existence, and, unfortunately, Jung himself conveys the impression to the general reader, although he probably does not maintain it in principle, that these archetypes are entities in themselves, and that they 'exist' in our 'collective unconscious'. Indeed he says that they embody an experience 'older than the individual'.

There seems no objection to our speaking thus of the archetypes, just as we speak of the 'will' or the 'conscience', provided we do not fall into the error of considering them as entities in themselves, and regard as objective realities what are in effect only abstractions or at most functions of the personality. For that reason it would be preferable to speak of these functions as 'archetypal', using the adjectival form to express the idea that they are 'archaic', rather than as 'archetypes', as though they represent things – for quality must not be confused with substance.

But whilst we must avoid the tendency to regard the archetypes as 'things' or as entities in themselves, *that does not prevent us from regarding them as realities*, which we assume is what Jung above all things wants to establish. The forces of evil in the world, so active, so potent, have for ages past been personalized as the Devil. We have ceased to believe in the personal Devil, and in doing so have perhaps lost something; for we have ceased to fear evil and it stalks unheeded through the world. The Devil, in fact, does not exist, but there is plenty of devilishness in the world today, which is the expression of evil forces in evil men. It is of course very effective to tilt at the Devil as a person and condemn not only him but all his works; but surely we can ultimately cope with these evil forces in the world much more effectively by seeing them to be what they are and dealing with them as the evil that is in men rather than personalizing them as the Devil. So we may vividly and effectively speak of the 'Archetypes' as arising from our 'collective unconscious', and as possessing great power in our personal as in our communal lives. But what in fact are they, how do they originate, are they inherited or are they derived from individual experiences of a common if not universal nature?

THE TRANSMISSION OF ARCHETYPES

The problem of the transmission of these potentialities is one which no one has yet solved. How far and in what sense are we to regard the archetypes, or for that matter any mental process, as inherited or racial? Jung says that every noble deed or act of devilry of our ancestry is contained in the unconscious. If by this he means that these deeds of our ancestry directly affect us hereditarily, it is difficult to believe, as it appears to contradict what we know of the laws of heredity, especially the non-inheritance of acquired characteristics; but if he means that the capacity for good and the capacity for evil are in the race, and therefore in every individual in the race, there is no need to quarrel with the idea. We have seen many a child capable of every deed of devilry, and no doubt its ancestors were capable of the same thing, but that does not necessarily mean that the child is so because of the deeds of its ancestors: it only means that we all have the capacity to do evil, whatever that is. So aggressiveness and even cruelty are natural reactions whereby an animal survives; their potentialities are inherited, not they themselves.

What we do inherit is a body with organs of circulation, respiration, digestion, a neuro-muscular and a glandular system. These organs and systems when stimulated give rise to functions and reactions which serve the survival of the organism; and some of them also give rise to mental processes, feelings and emotions such as fear and sex, serving the same purpose. The lactogenic hormones, for instance, when poured into the blood stream not only produce milk but also feelings of tenderness, and are probably the physiological basis of the so-called 'maternal instinct'. We do not inherit tenderness, but we do inherit the glands which secrete the hormones which produce tenderness. So it would seem to be with other instincts and perhaps other mental processes. If we ask, then, do we inherit mental characteristics, we may suggest as an answer that *in so far as our mental processes depend upon our physiological organism, so far we may be said, for practical purposes, to inherit them.* What is inherited, however, is not the mental process but the physiological conditions of their production. In this sense we may say with Jung that we do not inherit the archetypes, but only their potentialities.

If, then, we take the trouble to trace the individual archetypes back to their source, by reductive analysis, we may divide them into those which are thus dependent upon functions of the physiological organism; others of them originate in individual experiences in childhood; whereas others are the projection and objectification of abnormal organic processes, which often give rise to nightmares such as we shall later describe. All of these may appear in dramatized and personal forms in dreams, but their origins are different.

We may therefore distinguish various groups of these archetypes according to these sources.

(*a*) Some of these archetypes are probably innate potentialities in so far as they are dependent upon the functions of our physiological constitution and as such may be regarded as transmitted. That seems to be the case with the *anima*. As Jung has pointed out, we all inherit a physical constitution which is part male and part female, probably determined by our glandular system; and there is nothing to prevent our accepting the fact that these have their counterpart in our mental life and constitution. There is no harm, therefore, in calling these functions (in the male) the *anima*, as long as we regard this as a function or aspect of the personality and not as an entity in itself. The fact that the 'eternal feminine' is personalized by poets and in our dreams should not disguise from us the fact that the eternal feminine is only an abstraction. The personalization of concepts is poetic, it is not scientific.

What, however, is of importance is not whether the *anima* exists as an archetype, but the fact emphasized by Jung that the *anima* and other personalized and objectified creatures which we see in our dreams represent *qualities in ourselves,* in this case feminine qualities which exert an influence upon our personal lives and behaviour. In dream interpretation we translate these persons back into potentialities, these figures into functions; and this applies whether we regard them as derived from complexes originating in early childhood or further back from racial experiences and the racial unconscious.

(*b*) Archetypal forms like that of the *persona* are in a somewhat different category. This mask that we wear, this façade we assume before our fellows, is surely adopted under stress of social pressure and in response to what is demanded of us in childhood by parents and others. It may be based upon deep-seated needs in the child for protection and security, which compel the child to seek for the approval of those on whom it depends and with whom it identifies itself, and therefore to take over their moods, characteristics, ideas, and behaviour, which may be contrary to its own nature; in other words, to impose on itself a mask or *persona* more conformable to their wishes. This *persona* is false in so far as it is adopted at the expense of repressing the child's real self. The conflict between the child's natural personality and this fictitious *persona* is one which may lead to a neurosis, and the problems concerned in this conflict often emerge in dreams in which the *persona* is naturally represented as a person. The policeman in the dream on page 133 was such a *persona* with which the patient at first identified himself, but which in the series of dreams was gradually abandoned in favour of a more natural and healthy attitude of mind. This distinction is not merely an academic question, but is important practically; for if we regard the *persona* as something acquired in early childhood, we can set about finding out its cause and getting rid of it, as happened in the case of the policeman, whereas if it is something racial and archetypal we have to adjust ourselves to it as best we can. In brief, we regard the *persona* as belonging to the personal, not to the racial or collective unconscious, although the deeper needs which give rise to it, such as helplessness, the need for security, and therefore the need to adapt ourselves to others, may be regarded as innate, the actual adoption of the *persona* being made more or less necessary by circumstances. These things we may discover if we reductively analyse a patient's specific type of *persona* to its origin in childhood experiences, a process Jung has not allowed himself since he does not believe in searching for such causes and complexes.

(*c*) There are other archetypal forms which definitely arise from abnormal experiences of a more individual and personal kind. Take the instance of the 'Terrible Mother' or the 'Evil Eye'. It is of course possible that such ideas may be racial and may be derived from images in the collective unconscious. But if we take the trouble to analyse them out reductively we commonly find that such a dream as that of the 'Terrible Mother' relates to an experience with a mother who at one time was loving and tender and then suddenly became angry, glaring, and terrifying – how terrifying only a child who has experienced it can know. It is this same experience, so common to childhood, which is represented in the fairy story of Red Riding Hood. It also appears in some myths and fairy stories as the witch; and also in some religions as the 'cruel goddess'. These images we have frequently been able to trace back to the personal mother, for unkindness of a mother is a common although fortunately not a universal experience. Not all children have dreams of the 'terrible mother', because there are a few mothers who do not get angry with their children. We should therefore ascribe such dreams also to the personal, not to the racial experiences, to complexes, not to archetypes.

Similarly the idea of *guilt* which is so closely bound up with sacrifice, especially in religions, is not an innate quality but is nevertheless almost inevitable since the child lives in a social world which condemns those individual impulses which are disapproved by the community. This moral attitude of the community is adopted by the child and becomes its 'conscience' and gives rise to feelings of guilt. This guilt may be, and is, often repressed, and so becomes unconscious, but it is a quality which, as such, often emerges in dreams such as in the 'infanticide' and the 'apple' dream (pp. 26 and 53). Obviously such guilt belongs to the personal not to the racial unconscious: it is common to all individuals in all races and peoples, not because it is archetypal, but because it is a universal experience that every individual has to conform to social demands and is condemned if he does not.

(*d*) But most interesting of all are those dreams of vampires, spiders, snakes, crabs, and other monsters of the deep which appear to come up directly from our archaic ancestry but which on investigation turn out to be the representation, objectification, and personalization of abnormal organic sensations, such as appear so often in nightmares. We shall demonstrate these in the chapter on Nightmares (pp. 191 ff.).

Against this explanation, however, Jung brings forward evidence that there are images and pictorial representations, corresponding to myths, which come up in the dreams of his patients, which these patients could not possibly have experienced in life or heard of; therefore they must have emerged from the collective unconscious. That of course would be strong evidence of the racial nature of such images if it could be established, but it is very difficult to prove that those who have such dreams have never had access to a book of folk stories, seen such pictures, or heard tales of such nature. Moreover, we shall find in the chapter on Nightmares that many of these mythical monsters said to arise from archaic experiences *are* in fact derived from individual experiences such as an orgasm in an infant which is personalized and projected in these forms. This is therefore a subject on which we must suspend judgement.

But it is still open to doubt whether such personal experiences can cover *all* these bizarre notions and phantasmagoria of dreams; and there is no reason why *images* like 'the hero' cannot be derived from ancestral patterns of experience as plausibly as *patterns of behaviour* such as 'the *instincts*' (see p. 42). But the possibility that these images and ideas are derived from common individual experiences and needs of early childhood must be fully explored, for such is demanded by the principle of the 'parsimony of hypothesis', which declares that we should accept the simplest explanation which fits the facts; and no doubt in this case the simpler explanation is that such experiences and archetypes are derived from childhood – if they can adequately be so explained. This gives them a rational explanation and one which will enable us to cope adequately with them. Otherwise, if we delve too deeply into these so-called archaic experiences, and regard them too readily as mysteries of the racial unconscious, we may ourselves be overwhelmed by them and suffer the fate of one of whom it was said that he had locked himself up in a room full of archetypes and then could not find the key to get out!

Let us be reasonable even in considering mental phenomena which are themselves not reasonable.

PART TWO

THE BIOLOGICAL THEORY

CHAPTER 4

THE FUNCTION OF THE DREAM

ACCORDING to what we shall call the Biological Theory of dreams, the function of dreams is that by means of reproducing the unsolved experiences of life, they work towards a solution of these problems. It is not claimed that this theory is entirely new, as it is based on the theories already described, the popular as well as the scientific, but it gives to dreams a new orientation and in particular ascribes to them a more important biological significance than the earlier views provide.

DREAMS STAND FOR EXPERIENCE

Let us take the simplest kind of objective dream. A man who when climbing a cliff had had a slight slip, but had come to no harm, that night dreamt he was cliff-climbing, slipped, and had a terrifying fall as he hurtled to the ground, and woke up with a start. The next night he had the same dream, but in falling he tried to clutch a passing stump of a tree or jutting rock, and again woke with a start. On successive nights he dreamt the same thing, but ultimately he did catch the jutting rock and so saved himself. Now the effect of this series of dreams is that if he is subjected to such an experience again, which he is liable to have as a cliff-climber, he is better able to cope with the situation, for he has now been through the experience not once but a dozen times. Such repetitive dreams are very common to airmen who have crashed, to soldiers and civilians who have been blown up or buried, and who sometimes for months or years continue to have war dreams, and to most of us if we have even slight car accidents which might have had serious consequences.

These dreams clearly indicate the biological function of dreams. In the first place, *dreams stand in the place of experience*. Thus by making us relive the experiences and difficulties of the day in imagination they relieve us of the necessity of going through the actual experiences by trial and error and thus save us from many a disaster.

Secondly, by making us relive these experiences, often in an exaggerated form, they warn us of the possible consequences of our actions, and may indeed prevent us cliff-climbing by producing a phobia for cliffs, or deter us from ever taking the risk again by producing a 'nervous breakdown'.

Thirdly, by reliving these experiences dreams work towards a solution of the problems. That is the significance of the repetitive nature of the dream: the unsolved problem is perpetually thrust into consciousness until we attend to it and solve it.

Thus if a novice is sailing unwarily, has a sudden gybe and nearly capsizes, he may dream first of actually capsizing, and perhaps even drowning, for the first dream of such an event is usually an exaggeration, and he wakes up in a start. This exaggeration of the consequences of his action serves to warn him of its dangers even more than the experience itself. But as he continues to dream of the incident the next night and the next, he finds himself coping with the situation by righting the craft before it capsizes, either by turning into the wind or letting out the sheet gradually, and so he becomes more skilled in dealing with this problem the next time he meets it.

To take a more subjective problem: a man has trouble with his boss in business and he has to submit. This rankles in his mind, yet he cannot do otherwise than give in for fear of losing his job. But he has a dream in which he stands up to his boss, and to his surprise his boss respects him the more. This dream makes him realize that it is because he cringes that his boss despises him and that by having more confidence in himself he will win the respect of others. He previously felt he ought to do something about it, but did not. The dream, by making him actually do so, encouraged him in fact to take up that attitude with gratifying results. The dream stands in the place of experience and helps towards a solution of his problem.

In the pursuance of this aim and purpose, dreams have certain characteristics which we may study.

In the first place, dreams are essentially reproductive: this is the most popular and, so far, correct view of dreams. They may be direct and literal reproductions of events of the day, or they may be reiterations of events in our remote past such as terrifying experiences of infancy which often appear as nightmares and are reproduced as phobias. Or they may be reproductions not so much of events but of problems and worries which in the dream may be expressed in symbolic form.

If, however, we examine these experiences we shall find that they are rarely *exact* reproductions of what occurred during the day, nor are they by any means the most important events of the day: indeed they are often of the most trifling nature. Nor are they necessarily the events with the greatest emotional tone. But if we study them more closely we shall always find that they are experiences which are in some way connected with an unsolved problem or worry. The experiences of our childhood which we dream about are always those which we were unable to cope with at the time – for those fears and situations with which we were able to deal adequately leave no unsolved problem behind to worry us. These terrifying experiences of childhood persist and reproduce themselves throughout childhood until they may be solved by the greater confidence of adolescence, when they tend to pass away; or they may persist through life.

DREAMS AS PERSEVERATIVE TENDENCIES

This tendency of the mind to reproduce worrying situations is what psychologists call *perseveration*, which has been defined as 'the tendency of the mind to revert to an unfinished train of thought'. Dreams are a form of perseveration, for they reproduce over and over again the difficulties with which we have been confronted but not been able to face, problems we have encountered but not been able to solve, duties which were demanded of us but which we failed to fulfil.

When an organism is frustrated in its endeavour, whatever this may be, it does not immediately abandon it, but continually reverts to the problem, varying its behaviour so as to overcome the obstacle until the need is satisfied. The lower animals do this by trial and error so that the rat in a cage dashes about wildly until it finds a way out. Human beings do it more intelligently by deliberation and forethought. But when that proves hopeless and we abandon the effort, that does not necessarily put a stop to our trouble nor to the urge to solve it, and then we are said to worry.

Worry is indeed a typical form of perseveration met with in ordinary life. When we worry it is because we have some difficulty which, since we cannot attend to it at the time, or because we did not want to face up to it, or because we could not in any case solve it, we try to put out of our minds. But it will not allow us to ignore it; it insists on thrusting itself on us, comes back on us time and time again, intrudes upon what we are trying to attend to, so that we cannot concentrate on what we are trying to do throughout the day, and at times are nearly driven to distraction until we do attend to it.

Perhaps the most familiar example of perseveration is that which occurs after an unsuccessful personal encounter. We are, shall we say, the victim of some remark at which we take offence but to which on the spur of the moment we find no suitable retort. We dismiss the incident and try to forget it, but it rankles and we find ourselves spontaneously reliving the episode, but completing it in more appropriate ways, by some neat or cutting rejoinder. All this takes place not by any effort of will or by deliberately directing our thought to the subject, but automatically and spontaneously: indeed we may try to get rid of the thought from our minds, but it goes on irrespectively. Again, these perseverative trains of thought are not merely revivals of the past experience. They also have reference to the future and may in fact take the form of worries regarding possible future difficulties.

This is precisely what happens in the dream – we have worries and difficulties and may more or less succeed in pushing them out of our minds during the day; but at night when our will is in abeyance these worries all come rushing back into our minds and have a regular field-day, so that even if we do not have definite dreams we have a disturbed and restless night, and may be more tired in the morning than when we went to bed at night. At other times we may succeed in dismissing the whole thing from our minds during the day and forget all about it. But even so the perseveration persists, and what we refuse to think of during the day emerges in sleep in the form of dreams. Dreams are revivals of past unsolved problems, and as such have the character of automatically attempting a solution in imagery. Thus they may influence our subsequent behaviour, and even change the course of our lives, as in the case of some visionaries and revolutionaries. It is this process of perseveration which gives to dreams their repetitive nature, and their problem-solving function.

DREAMS AS A FORM OF IDEATION

It is obvious, therefore, that *dreams serve the same purpose as ideational processes*, much as we exercise in normal thought in waking life. Ideation is the power we have of forming images and ideas in the mind in the absence of the actual experiences themselves. If these images or trains of thought are concerned with the past, we call them memory; if with things which have not yet happened, we call them imagination. Now this process of ideation especially imagination, whether in waking life or in sleep, enables us to work out our problems, whether of the past or the future, without the necessity of going through the actual experience. If we plan a holiday on the continent, we do not vaguely start out for the coast on the off-chance of getting there by the process of trial and error; if we did, we should be very unlikely to arrive. We think things out beforehand, go through the journey in imagination, decide what we are likely to need, visualize possible dangers we are likely to have to face, difficulties we are likely to encounter. We recall past experiences to help us to avoid the mistakes we made last time, and so remember to take our passport and some foreign exchange. All this we can do from our armchairs at home with the map before us and the telephone at our side. The most effective way of establishing a new skill, such as learning to swim or to dance, or, if we are going on the Continent, learning to drive on the right side instead of the left, is to *imagine* ourselves doing it over and over again, with all the movements, all the hazards, and all the difficulties we are likely to encounter. It is the same with any sport and gives rise to the saying that we learn to skate in summer and swim in winter. The actual practice which is necessary to learn the sport can be greatly reinforced by practice-imagination.

Thus ideation helps us to solve our present problems; it also enables us to plan for future action.

This power of the mind to reproduce in imagery the experience of the past and possibilities of the future lifts the life of the human being on to an entirely different plane from the life of the lower animals; for it is no longer necessary for us to go through an actual experience with all its hazards and risks of disaster before discovering the result and consequences of any particular line of action; we visualize much of it beforehand, and can thus forestall them. Thus those animals like man who have the power of forming images and living through such experiences in imagination have the advantage over those who have not the capacity of ideation, but are restricted to the functions of perception only.

Dreams are a form, probably the most primitive form, of ideation in which experiences and situations of the day and of life are reproduced on the screen of the mind during sleep as images, usually in visual form. By the mere fact of reproducing the experience they enable us to live through the experiences and so work out our problems to a solution. It is said that the hound relives the experiences of the chase while it sleeps before the log fire and gives an occasional bark. If so, we hazard the guess that the hound is dreaming of the hare that it did *not* catch! It is probably reliving that experience, and by doing so it will make sure that the hare does not get away next time!

Dreams, like ideation, can not only reproduce unresolved problems of the past, but can also anticipate the possible consequences of our actions. Dreams, like ideation, relieve us of the necessity of perpetual efforts of trial and error and by enabling us to foresee the likely consequences of our errors help us to avoid them. A lady was driving a car, and, feeling for her handkerchief, changed her right hand to the left side of the steering wheel. She swerved, ran on to the path on the country road, and narrowly missed landing in the ditch, but righted herself in time. After a moment's fright she made light of it, no doubt to excuse her foolishness. That night she dreamt of the car swerving, but in the dream it drives across the road, down the steep bank beyond, and is completely wrecked. That was not what happened but what might have happened. Moreover, the dream enforced this on her the more vividly by making her relive it. In the day she makes light of it because she dislikes being reminded of her foolishness: her dream, working by the process of perseveration and ideational reconstruction, reverts to the incident and compels her to face up to the situation by emphasizing the possible consequences of her carelessness – an idea she had conveniently repressed. The terror of this dream had the effect of making her much more cautious, whereas the experience of the day failed to do so: it was a corrective to her false optimism.

We see, then, that in reproducing these situations dreams are fulfilling precisely the same biological function as ordinary ideation; they stand in the place of experience, they recall unresolved problems and difficulties, they warn us of the possible consequences of our actions, and so help us to solve the unsolved problems of life. They are helped in this task both by memory of the past and imagination for the future.

How, then, do dreams differ from ideational processes of thought in normal waking life? The difference between ideation and dreams is that whilst the former is deliberate, rational, and logical, dreams are spontaneous, automatic, and dramatic. Ideation or thinking things out has the advantage over dreams in that by such means we can work out by the cold light of reason the logical chain of cause and effect with much greater accuracy and in much greater detail. But what dreams lack in precision they make up in vivid representation.

Thus in some respects dreams can perform these functions of solving our problems even more effectively than deliberate thinking or reason. For one thing, reason can only think out problems of which we are consciously aware, whereas in our dreams there are revived in our minds those problems which the conscious waking mind has deliberately discarded, and of which we are now unconscious but which nevertheless affect our lives adversely and press for solution. We cannot think out problems which we do not recognize as existing; the dream vividly recalls them to our memory. But this is not all: the dream has a different method of solving the problem from that of conscious reasoning. For these problems have been banished from the conscious mind to the subconscious; and the subconscious mind works by methods altogether different from those of the conscious mind, as we shall see later, the results of these conclusions being given in dreams. Thirdly, though, as we have said, ideation can present a problem much more explicitly, dreams have the compensating advantage of visual dramatization – they reproduce the feeling of reality characteristic of actual perception, together with the emotional vividness that this entails. Thus dreams are more capable of solving our problems by virtue of two characteristics, not shared to the same extent by ordinary ideation, namely the *exaggerated emotion* in the dream and the *dramatization* of dreams, the fact that we actually live in the dream and are actually partakers of it. Let us illustrate these. They are rarely simple reproductions.

EXAGGERATION IN DREAMS

The emotion of the dream is usually exaggerated far beyond the experience of the day, so that it produces a deeper impression upon us than the actual event and urges us to more immediate action. This was illustrated in the case of the cliff-climber and the lady who swerved in the car, the consequences of which in the dream were far more disastrous than the experiences of the day. A boy of nine, angry with his brother, threw a knife at him, and narrowly missed him. He dreams of the experience again, but in the dream the knife strikes and kills his brother. Then in the dream he has the horrible experience of being accused as a murderer, condemned, and hanged – which has such an effect that he thereafter represses all his aggressive tendencies. Such dreams are both reminiscent and premonitory; they not only remind us of what did happen but of what might have happened, and that so vividly and dramatically that they serve the most useful biological function of bringing home to us the possible consequences of our actions. For the terror this boy felt at the time he threw the knife was nothing to the terror he felt in his dream.

This exaggeration in emotional tone in the dream may be explained physiologically by the fact that in sleep the inhibitions of the cortex of the brain are cut off so that the emotional centres are unrestrained, and thus are far more compelling and vivid. However that may be, the *effect* of this exaggeration is profound, for the exaggeration of the fear or guilt in dreams may have a more telling effect on our life than the incident itself; and may produce a greater repression of the emotions than the event which gave rise to it. Thus we often find in treating the neuroses that the repression of the emotion which produced the neurosis was the result of some terrifying dream or nightmare, and in our analysis we often make a point of looking for such a cause. In this particular case the dream was too successful, for his terror was so great that he repressed all his aggressiveness and became a docile boy, with the result that the fear of his aggressiveness later emerged in the form of an anxiety neurosis, which represented his fear of his aggressiveness. Nightmares, as we shall see, have a particularly profound effect, so that if, shall we say, sex feelings are associated with a terrifying nightmare of vampires, that is enough to produce a repression of sex feelings with possible disastrous effect upon the later sexual and marital life of the person involved.[1]

This exaggerated effect of the dream is further accentuated by the fact already illustrated that in dreams we dramatize, actually live through, and are partakers of the exaggerated experience. In the day the boy who threw the knife no doubt knew that if he killed his brother he would be charged with and perhaps be found guilty of murder: but he had *not* killed his brother, so what? Beneath the surface of consciousness, however, or, as we say, at the back of his mind, he is not altogether satisfied with that complacency, for he might do it again with more disastrous consequences. The problem still worries him and recurs in the dream in which he imagines that he does kill his brother, is tried for murder, and is hanged. It is no longer a question of *if* you do this, such and such will happen; he *does* do it and he actually experiences the consequences – to his great horror.

The girl who is jilted in the day saves her pride by a pretended indifference and says 'Who cares?' But in her dream she lives through the feelings of self-pity she has repressed, and wakes up weeping and in misery. Thus what she refused to face is forced upon her waking consciousness and compels her to admit the distress which she can hardly now deny. So the bumptious individual whose arrogant manner gives him an air of superiority to his fellows has a dream in which he is being laughed at and scorned by his circle of friends. This experience, though only a dream, has such an unpleasant effect on him that he is more subdued in meeting them the next day and is less likely to make a bumptious fool of himself.

Many dreams relate to horrifying experiences of the remote past, even to infancy and birth itself, which leave behind them problems which the child, being helpless, finds it impossible to

1. See *Psychology and Mental Health*, p. 275.

solve. These traumatic experiences of childhood often become repetitive and nightmarish and, like so many nightmares, repeat themselves over and over again until they are resolved. A dream of this nature is the following:

A patient had a dream that 'he is in a laboratory and there is danger of an explosion. So he goes on hands and knees and passes out of the laboratory. But none of the others mind.' Taken symbolically, that explosion might mean all kinds of things, such as the explosion of repressed emotions; but with him it went back to an incident he was analysing when a nurse in anger put a pillow over his head, which he felt getting fuller and fuller till he was obsessed with the dread that his head would burst: but (as in the dream) his knees gave way and he 'passed out' as his head 'exploded'. The lack of concern on the part of the other people might signify two things, the indifference of his mother to the nurse's cruelty, which negligence in fact impressed him deeply, *or* a reassurance that his fear was not as bad as he thought and he must not take it too seriously. Probably it was a case of condensation and referred to both. One of the neurotic symptoms from which this patient suffered was the feeling of extreme tension and pressure in the head, derived from the same experience. The dream was trying to bring the experience to the surface of consciousness for attention and solution, and the analysis, following the same path, succeeded in doing this so that he was cured of his symptoms.

The following dream of the same patient was subsequent to the dream above. He dreamed that he had murdered a policeman, and he knew that the police never forgot if one of their number was murdered and would pursue him to the death. The policeman might be taken endopsychically to represent his conscience pursuing him (as in the dream on p. 133 f.). But to him the nurse was the policeman, and he himself related the dream to the fact that after the incident with the nurse he felt furious with her and would have liked to smash her up and trample on her dead body; but he dared not, for she had just nearly killed him and there is no knowing what she might do to him if he attacked her – she would pursue him to the death, as in the dream. This dream was not to be interpreted in any complicated symbolic way, for it related to facts and to terrifying experiences which left behind them residual moral problems.

SUBJECTIVE PROBLEMS IN DREAMS

But, as no doubt the reader has been objecting, the majority of the problems of our lives are not objective difficulties like climbing cliffs or driving cars or throwing knives: they are problems within ourselves, the dangers which arise from our impulses, our aggressiveness, our sex passions, the problems of jealousy, of humiliation and insult, of moral failure, of defeat, of a sense of inferiority, or self-consciousness.

But precisely the same principle holds good with these subjective as with the objective problems of our lives: in dreams we relive these problems, and in reliving them we are helped to solve them. But since the subconscious mind cannot express itself in abstract terms, such dreams are put in symbolic language.

Let us take the simplest imaginable dream of this nature. A man simply dreamt, 'I saw a hand painting a clean, neat band along the keel of a canoe.' According to the Freudian view that would be a phallic symbol. On the purely reminiscent theory the dream can be explained by the fact that on the previous day he was painting with bitumen the keel of a canoe he had built, since it leaked. But why should he dream of so unimportant an incident? Nor was it purely reproductive, for the band he had actually painted was by no means straight and neat as the one in the dream, and seeing the straight line in the dream immediately reminded him of this fact. Indeed the fact was that he was dissatisfied with his work at the time, but persuaded himself that 'it would do'. In brief, he made a shoddy job of it and knew it, but pushed the matter aside because he could not be bothered. But the dissatisfaction with his work must have persisted and worried him subconsciously, for the dream would not let the matter rest. It reproved him by producing a picture of a perfectly painted band; in other words, it said, 'That is how you should have done the job!' – which of course he knew and had told himself, but had ignored. Even so, that might be considered a small matter to worry about and dream about. Not at all! It was not such a trifling matter as at first appeared, for the dream not only referred to this particular job shoddily done, but to a defect in his character which needed rectification, namely the tendency to make a thing do instead of making a thorough job of it. The dream was in fact a revelation of his slipshod methods of working.

DREAMS NOT MERELY REPRODUCTIVE

Dreams, as we have said, are reproductive, and a superficial explanation would be satisfied with that. But, as in the case of the canoe and most other dreams cited, it was not a simple reproduction; indeed in the canoe dream the line drawn was the opposite from the fact. Why was the dream so different? The reason is, of course, that the dream is dealing with a problem and trying to work out a problem and is therefore bound to change in attempting and suggesting a solution. It demonstrated not what he had done but what he ought to have done. Indeed it aroused his conscience and worried him to such an extent that the next day he did the job properly. Further, this experience of the canoe was chosen as the subject of the dream out of thousands of events of the day because it related to a deep subjective and moral problem of which this incident was but one instance.

This selection of an instance to illustrate a principle very commonly happens in dreams, and explains why it is that we dream of apparently trifling incidents of the day, or recall experiences of the distant past which we have long since forgotten. For these experiences, as we have said, are not as a rule the most important events of the day, nor indeed are they necessarily those most charged with emotional tone. But the analysis of such dreams always reveals that these experiences contain some problem which has been left unsolved and which has perhaps been awakened by some present-day situation.

The dream of Largs (p. 10) is a typical instance of the way in which an old, apparently indifferent memory recurs, because it is associated with a present-day problem. The sudden glimpse of a scene in school-days may recur, we do not know why. If we investigate it more deeply, we find that it refers to some unpleasant experience, shall we say, when we were laughed at for some characteristic which we had then but which we still have, and which for some reason has recently been revived. The face in the crowd which comes back to us may remind us of someone in the past to whom we owe an unpaid debt of gratitude. The trifling story we read in the evening paper but which keeps recurring to our minds or in dreams probably has some personal significance. Whilst dreams therefore in general are fundamentally reproductive, they are rarely exact reproductions of the events of the day: either the facts are different or the emotional tone is altered – we were afraid in the day, we are bold in the dream. This is obvious in other dreams already mentioned; for instance, the canoe dream, the dream of the lady driver, or of the man who in his dream stood up to his boss instead of cringing. This distortion is not necessarily due to an effort of the pre-conscious to disguise an unconscious wish. It is because the function of the dream is in fact first to call attention to the problem and then to suggest a change in our attitude of mind towards the experience. So if we have been too unkind and cruel to someone but silence our conscience, we dream of the event, but in the dream are moved to sympathy. If the dream is working out a problem or making tentative suggestions to that end, it is bound to change and must alter the facts to suggest the solution.

Therefore if we watch a person's dreams over a period, especially in analysis, we shall notice a progressive change in the nature of the dream, instances of which we shall give later (p. 131). This progressive change is itself significant as showing the psychophysician what progress the patient is making towards a solution of his neurotic problem, what changes are taking place in his personality, what efforts, successful or unsuccessful, he is making to readjust himself to the situation, and what solution is being arrived at. Sometimes, however, dreams are almost pure reproductions, often of experiences of the long-distant past, such as terrifying noises, of suffocation, of glaring eyes, of being abandoned and lost, and these traumatic experiences and fears of early childhood are sometimes themselves sufficient to create insoluble problems of fear of facing life which are perpetuated in dreams as well as in neuroses, and repeat themselves over and over again as nightmares (p. 175 f.). Thus the child who is 'from his mother's womb untimely ripped' suffers from the terror of separation and may later suffer dreams of being lost, 'separation anxiety', and a neurosis which takes the form of a fear of leaving home or travelling by train. Such repetitive dreams serve their function when they call attention to the ignored problem, without offering any solution. The dream of 'Largs' is a case in point.

In other cases the psychological problem has to resort to an entirely imaginary situation in order to represent a present-day problem. A married woman whose husband has left her tries to save her pride by bluffing herself that it was all the other woman's doing, and that her husband was deeply devoted to herself but had been led astray by this scheming harridan. She has a dream, however, in which 'I see myself travelling affectionately with my husband in a train, and then going along the corridor he meets the other woman, turns to me and says, "I am going off with this other woman".' This dream is not merely a wish-fulfilment that she is 'travelling affectionately' with her husband, it is compelling the patient to face a situation that she persists in avoiding, namely that it was her husband's own choice that he left her and that he did in fact prefer the other woman. It was a problem of her pride; and her subconscious mind refused to allow her to take refuge in the subterfuge and bluff.

This dream illustrates another important function of dreams. It is said that dreams, being bizarre and imaginary, have no relation to truth. The oppositei s the case; the dream, as in the case quoted, forced the truth upon her when she had refused to face unpleasant facts. Dreams are bizarre because of their symbolism, but they nevertheless deal with truth.

In all these instances, then, dreams, by repeating in imagery the problems and difficulties of the day, tend towards a solution of those problems. Thus dreams have a significance for life and the problems of living. That is why we call ours the 'biological' theory – using the term 'biological' somewhat arbitrarily, as signifying 'adaptation to life' – to subjective as well as to objective life.

THE BIOLOGICAL COMPARED WITH THE FREUDIAN THEORY

This theory gives to dreams a far more important biological significance than that of Freud, who maintains that the function of dreams is, by gratifying our repressed desires, to relieve the tension and allow us to sleep. The fact is that in many cases dreams do no such thing. Dreams are far more likely to *disturb our sleep* than to encourage it; indeed Freud admits that the majority of dreams are disagreeable in spite of his wish-fulfilment theory. The reason why dreams are disturbing is not because their wishes are forbidden by the pre-conscious, *it is because they are concerned with disturbing and distressing problems*, and to be so, in our view, is their function. Far from allowing us to sleep, as often as not they will not allow us to sleep, for it is more urgent that we solve these problems than that we sleep. We have all had experience of not being able to get to sleep at night because of some worry on our mind which demands attention, and which we must settle before we can relax enough to sleep.

Again Freud has recognized the repetitive nature of some traumatic experiences like that of war-shock in dreams, but apparently without recognizing that such repetition or perseveration is characteristic of all dreams; nor apparently does he recognize the biological significance of such repetition as a persistent means of thrusting an unsolved problem into consciousness for solution.

Again, Freud says that dreams by gratifying the wish release emotion and therefore relieve tension. This may be so in purely wish-fulfilment dreams. But far more frequently the dream does no such thing, for instead of relieving the tension, the terror in the dream *greatly increases the tension*, a tension which may be felt all through the subsequent day. The woman driver (p. 70) was comparatively lighthearted after her experience, for after all nothing serious had happened. The dream does not allow of such complacency and fills her with apprehension about the possible consequences which she was inclined to overlook, and in fact this added apprehension made her more cautious in future. The dream increased, it did not diminish, the tension, and it is this increase of tension which has so marked an effect upon our lives, as we have seen.

Again Freud says that the dream is wish-fulfilment, whereas in fact the dream often has the opposite effect of *repressing the wish*. In the case of the boy who threw the knife, the conscious wish was to hurt his brother, and in the dream he did kill his brother; so far you might interpret that dream as wish-fulfilment. But that is not the end, nor the point, of the dream, which emphasized the appalling consequences of his act to such an extent that it in fact compelled him thereafter to repress his aggressive impulses even more effectively than the actual happenings of the day. That, according to our theory, was the intent and purpose of the dream; indeed, as we have seen, it went too far and repressed them so effectively that he became a docile ingratiating boy and later suffered from an anxiety neurosis – as though to say, 'Better have a neurosis than be guilty of murder.' A neurosis in fact often prevents us from gratifying a wish and guards us against its consequences, as in the case of the man who planned to embezzle some money and go abroad, but thereupon developed a fear of travelling by sea.

Freud later on came near to what we have called the biological view when he said 'unsolved problems, overwhelming impressions continue the activity of our thoughts during sleep, maintaining psychic processes in the system which we have termed the pre-conscious.'[1] But this is apparently by the way.

SOMNAMBULISM

Somnambulism or sleep-walking is another illustration of the same principle. In somnambulism the sleep-walker is dreaming, but whereas in most dreams we only imagine ourselves acting in certain ways, in sleep-walking we make a move towards actually doing something about it. The analysis of somnambulist states reveals that the sleeper is invariably working out some problem. For instance, one child, badly threatened by her father, feels she must escape from the intolerable situation, but, being a child, she cannot escape and so has to suppress the desire. In her sleep she not only dreams of escaping, but she actually gets up and goes downstairs to escape, so urgent is the need. There she comes up against a blank wall and wakes unaware of what she was doing or why she was there. The situation was not eased by the fact that it was always her father who fetched her back. The somnambulism of children should always be taken seriously, since it always represents a problem or difficulty confronting the child. A married woman started sleep-walking in adult life. It originated when her father was seriously ill and she had, every now and then during the night, to go to see that he was all right. He died, but she continued the habit as a somnambulism. Her problem was not merely a perpetuation of the habit of caring for a sick father. It was a conflict in her mind: she did not want to lose him, but she had known his death would mean the end of her husband's financial difficulties. One motive was to attend to the sick father, the other to see if he was dead yet! Lady Macbeth's somnambulism is, of course, classical – an attempt to wipe away the stain of the guilt she refused to acknowledge.

1. *The Interpretation of Dreams*, 1931 ed., p. 310.

THE PATIENT'S OWN ASSOCIATIONS NECESSARY

In the analysis of dreams, whether of actual experience or of symbolic representation, it is necessary to get the patient's own associations if we are to avoid an arbitrary interpretation. It is true that there are dreams which are common to man and which have universal application. But most dreams are personal and have reference to individual experiences and personal problems in the patient's life, and these cannot be interpreted without our having the associations of the patient on those items, as well as knowing something of the patient's life and psychology. It is not for the analyst to give an arbitrary interpretation which may be based on his and not on the patient's associations, for even mythological symbols like that of the snake or water do not always mean the same thing. As a case in point, a patient had a dream in which the word 'hide' came. The analyst interpreted this as being that he had something to hide. No doubt he had! Who of us has not? But the patient's own associations (for which he was not asked!) were that it referred to the fact that he had previously been jilted by a girl in Hyde Park. Possibly both meanings were correct, and possibly the dream with its condensation referred to both, but one cannot ignore the patient's interpretation and association, for they related to a real emotional problem. That does not mean, however, that the patient's associations or interpretations are always to be accepted, for they may still be a means of side-tracking the real issue.

If we get the associations of the patient, dreams which appear on the surface to be pure nonsense turn out to have a most intelligible meaning, as in the following dream: 'I am standing before my headmaster; and throwing up my arms I cry, "I will be free, I will be myself", and the master with scornful smile replied, "Potato!" ' Sheer nonsense of course! But when asked for his association on 'Potato', he replied, 'Oh, obviously "peut-être" ' – an excellent example of association by similarity. Thus when he says 'I will be free', the headmaster scornfully replies 'Perhaps!' – or as an American might say, 'Oh, yeah!' or, 'So what?'

The patient, the youngest child in a large family, lost his mother at the age of nine, was bullied by his harassed father, and scornfully treated by his brothers as being a 'sissy'. He grew up with an inferiority complex and completely lacking in self-confidence. This dream revealed what he was loath to recall, his father's scorn of him. Yet the dream was not merely a repetition of his father's scorn, but, like so many dreams, related to an endopsychic conflict, that is to say to a conflict *within himself*. For as his brothers and father had scorned him, so he came to scorn himself, and that was the cause of his futility. He therefore was bound to be futile, for he had no confidence in himself; he never gave himself a chance. The 'headmaster' stands for his father, but also for that part in himself which was identified with and took over the scornful attitude of his father towards himself; it was the father in himself. In other words, the dream indicates that the cause of his futility was that whenever his real self wanted to be free to be himself, something else in him replied 'Potato!' – 'Perhaps! What is the use of your trying to do anything?' He came to realize that it was this *self*-depreciation which was at the bottom of his lack of confidence, and, recognizing its source, he was able to throw it off by reacting defiantly to the scorn of his family.

Or take this dream: 'I was in a road and a man was trying to give an explanation of something, but somehow I could not be convinced that the explanation was adequate. Then I went indoors into a cellar and there another man was putting an injection of sodium into the sole of my foot. The idea was as a test to see whether I was tough, and I was confident that I could stand the test.' Such a dream is impossible to interpret and sounds nonsense until we regard each part of it as a picture of a representing or separate idea.

The patient's first association was that he had been for several years considering the question of having psychological treatment. But he had been trying to convince himself (the man in the road, like most imaginary persons in a dream, was part of himself) that the cause of his trouble was something quite superficial and that he did not need it. But somehow he could not be satisfied with his explanation of his neurosis and ultimately came for treatment. This is represented by his going into the house (for treatment), and into the cellar (analysis into the depths of the unconscious). The patient was a timid self-conscious man, unable to stand up to life, so that in fact he did only three days' work a week and for the rest pottered about in his garden. His problem was how to get self-confidence. In the beginning of the dream lies his problem; he is trying to find an explanation of his neurosis and trying to convince himself it was something trifling, but ultimately had to seek help. In the cellar (the analysis) he was subjected to an injection in the sole of his foot to test whether he was tough: in other words, to see (according to his associations) if his soul (sole) was tough. But why sodium? This related to a time at school when he was sent out of chemistry class for fooling and then came to the window of the classroom, got another boy to pass him out some sodium which he put in the master's lavatory and caused an explosion. For once he was really daring and tough! The dream recalls this, and in testing his sole with the sodium in effect says, 'You remember that time at school when you really were daring and defiant. We shall see whether you are as tough as you were then.' The injection of sodium means putting some pep into him. In the dream he is confident as to the result of the test.

Then the dream goes on: the other man in the dream (really himself) goes out of the room to change his clothing – a very common symbol of the change of one's personality (*persona* = mask; 'clothes make the man', etc.) – and he comes back in clothes having a shirt with a stiff front. In short, instead of his weak vacillating personality he can now put up a 'stiff front' before the world. Dreams often make poor puns, like this one and sole for soul, because of their working by similars.[1]

All the characteristics of a dream are here portrayed: the problem is stated at the beginning, that of his anxiety neurosis and his attempts to get a plausible explanation of it, with which, however, in his dream he is not satisfied; the process of analysis in the 'cellar', and his being put to the test, and the solution of the problem in the end – confidence that analysis will enable him to recover the spirit of adventure he once had, and put on a stiff front to the world so that he can face life confidently – an encouraging dream!

1. A girl who had spent the afternoon serving ice-creams at a garden fête talked in her sleep and said, 'No, I have no more cornets (a form of ice-cream), but I can let you have a trombone!' This is a fairly obvious sex dream, the in and out movement of the trombone being an apt symbol of sex intercourse. (In writing this, my subconscious mind threw up an appalling pun and I nearly wrote, 'In this way the fête determined her fate!' Perish the thought!)

These cases indicate how dreams may appear to consist of a number of odd bits and pieces of past experience and bizarre imaginations, like the bits of a jig-saw puzzle, until we get the personal associations of the patient, in which case everything fits together.

It is therefore of the greatest importance that we should get the dreamer's personal association and not too readily interpret all dreams over-symbolically: far more of them relate to unresolved personal experiences of the past than we are apt to imagine. Only if we analyse out these dreams reductively, that is to say trace them back to their origins, can we appreciate their true significance and the importance of early experiences in determining our present-day psychological problems. Dreams relate to both. The neuroses, Jung has said, are due to present-day moral problems (p. 55). We agree, but contend that most of these problems would not have occurred at the present day were it not for such early predisposing experiences; and it is these experiences, which are causal factors in producing the neuroses, that so often appear in our dreams. By getting the patient's own associations on the dream we can unearth not only the present-day problems but these early experiences which caused them, and so radically cure the patient. This method we call 'Direct Reductive Analysis'.[1]

The dreams already given perhaps err on the side of simplicity, which is intentional, since I wished to make the issue clear. (It must also be borne in mind that the dream when interpreted sounds much simpler!) No doubt the reader has been saying that his dreams are far more bizarre, they are not so clear as all that in fact they are, as a rule, just a jumble of nonsense; nobody could make anything of them. It is true most dreams are much more complicated; indeed it would be possible to write a book of this size on the interpretation of one dream alone with all its personal and mythical implications.[2] But precisely the same principles apply to those complicated dreams as to the more straightforward ones given – for all of them represent unresolved problems, many of them give indications of their cause by reference to early experiences, in most of them the apparent nonsense may be

1. I have given a description of this method in *Psychology and Mental Health*, Chap. xvi, p. 410.

2. See the Miller dreams in Jung's *Psychology of the Unconscious.*

turned into sense by getting the patient's associations, and some of them have been found actually to suggest a solution of those problems.

Apart from these dreams which have a strictly personal significance, there are dreams which are 'racial' or 'collective', if by those terms we simply mean that they deal with problems which are common to humanity – problems such as the conflict between the desires of the individual and the demands of the community, the problem of what we are to do with our primitive impulses and how to control them, problems of guilt and of rebellion against authority. When such problems appear in dreams they are therefore expressed in symbols also common to humanity, and, as Jung has shown, take much the same form as myths, which also are expressive of these problems. In such cases there is some justification in translating the symbols somewhat arbitrarily, as long as we realize they may have a more personal association.

By way of illustration we may take a dream dealing with the basic problem of the individual's rebellion against authority, and the problem of how to cope with his primitive aggressive emotions.

A DREAM VARIOUSLY INTERPRETED

'I dreamt that I was staying in a country house, and after everyone had gone to bed I went downstairs to the sitting-room to get the coal that was left on the sitting-room fire to take to my own bedroom. When I had taken the fire and reached the passage outside, I was met by a Negro, primitive in race and primitive in type, who threatened me. I tackled him and got him down, but then did not know what to do next. Then came a female form and said, "Don't kill him, and don't hurt him, but send him to a reformatory." '

Such a dream represents a problem common to humanity – the rebellion of the individual against the community. We should interpret the dream thus: The dreamer stole the fire which belonged to the community for his own ends. But this act of defiance against society immediately roused in him all his rebellious forces, his primitive instincts and emotions, as represented by the Negro, which were a threat and a danger to him and which he had the greatest difficulty in repressing. But the repression of our primitive selves is no satisfactory solution, for it simply produces neurosis as it had done in his case. What then are we to do with these primitive forces? Are we doomed to be either immoral or neurotic? Have we to choose between rebellion and illness? The answer is given in the dream; the female form represented his intuition, and says, 'Don't destroy or repress these primitive forces in yourself, but reform and transform them to the uses and purpose of the personality.' This is a neat solution of this basic problem in life: we need neither rebel nor repress, but by transforming and redirecting our aggressive instincts we develop a strong will and character.

But obviously other interpretations could be put upon this dream.

On the 'personal reminiscence' theory, the dream might be explained as being due to the dreamer having, in fact, thought of getting some of the coal from the common room for his own fire the evening before, but had resisted the temptation. But why should we dream of this of all the experiences of the day? Obviously the dream simply took this experience, as it did in the canoe dream, to point to a problem of a far deeper personal significance.

According to the Freudian view this would be a sexual dream – fire representing the forbidden passion which the dreamer illicitly steals. The Negro would probably represent the father and the female form the mother, revealing an incestuous situation which calls for the struggle between the father against the son. The defeat of the Negro by the dreamer probably represents a wish-fulfilment of the dream to get rid of his father in order to possess his mother. As to what the 'mother's' final remark about the reformatory could mean we can only hazard a guess: it may represent the activity of the pre-conscious presenting a conciliatory front to tone down the sense of guilt; or it might be a suggestion of the 'mother', whereby he might get rid of the father without incurring the responsibility of murdering him.

An Adlerian could probably interpret the dream, also with considerable justification, as the urge to power, the dominance of the masculine urge as against the inferior feminine role. In his later teaching, in which lack of cooperation became the keynote Adler might have regarded the dream as indicating non-co-operation as represented by stealing the fire and the trouble into which one is led by such non-cooperation and pampered attitudes towards life.

The Jungian interpretation might be much nearer the one we have suggested. He might point out that this dream resembles the myth of Prometheus, who stole fire from the gods, an act of defiance which landed him in all kinds of trouble. The Negro represents the powerful forces in the Collective unconscious which when they emerge into consciousness are liable to cause trouble. The female form might be the *anima*, or feminine function in the personality, to which the dreamer requires to give expression in order to compensate for those powerful forces. The teleological and purposive aspect of the dream is indicated in the final instructions of the *anima*.

But one outstanding feature common to all these interpretations is that in each and all of them there is the presentation of a problem, whether how to solve his Oedipus complex (Freud), his power urge (Adler), or how to integrate the forces of his racial unconscious (Jung). In each case the dream deals with an unsolved problem, and in whichever way we interpret it there is a movement to solve that problem, whether we regard it as a wish-fulfilment, a compensating power urge, or as dealing with a more deep-rooted moral problem.

Dreams dealing with subjective and moral problems are just as repetitive as objective dreams, but it is the problem, not an event, that repeats itself, sometimes in the same form as in repetitive dreams like the one of trying to catch the train and missing it; or it may be in many different and changing forms, just as when we find it difficult to convince a man by an argument we say, 'Well, look at it this way!' So our dreams, in trying to bring up problems relating to our primitive emotions, our feelings of inferiority, our conceit, or ingratiation, or cynicism, or selfishness, or our clinging to security, present the same problem night after night, in the same form or by different stories, but always with the same purpose – to bring the problem to a head in consciousness so as to force us to attend to it and solve it.

But not all dreams arrive at such a satisfactory solution, which is understandable when we remember that dreams only represent the *effort* of the subconscious mind to work towards a solution The complete dream does in fact achieve this purpose, but many other dreams fall short of completeness. Some dreams get no farther than merely thrusting the problem before us; some offer a partial solution, some an unsatisfactory solution, whilst others point the way to a true solution.

STAGES IN THE SOLUTION OF DREAM PROBLEMS

We may therefore distinguish several types of dream according to the success or otherwise of their efforts and the degree in which they manage to achieve their end.

(A) Some dreams, failing to find any way out of the difficulty presented to them, 'solve' the problem by the simple device of imagining it solved – this is the simplest but a most inadequate way of solving a problem – by wish-fulfilment.

An illustration of such wish-fulfilment is the dream already mentioned of the man who hears the alarm clock and dreams of church bells, or that he is now at work. That contains the simple problem, that he should be at work but wants to sleep. He 'solves' the problem either by saying it is not the alarm but church bells, or by saying that it is all right, he *is* at work. He gets out of the difficulty by imagining it solved. That corresponds to Freud's theory – it removes tension and allows the dreamer to sleep. Or again, a child needs to get up in the night to pass water, but he is too lazy and comfortable, and can't be bothered; so he dreams that he gets up, passes water, and gets back to bed, only to find to his shame in the morning that he has wet his bed. It is mostly in children that dreams take the form of simple wish-fulfilments, and that for three reasons: because children are so much frustrated in their wishes; secondly, they have so few resources for the attainment of these desires; and in the third place, and as a result of these, because a child's life is so much a matter of fantasy and day-dreaming anyway.

The dreams of children are therefore, as Freud has said, most commonly pure wish-fulfilments. But for him to add 'They present no problem to be solved' is far from correct. As an instance, he takes the dream of his son that he had climbed to the Simony Hut.[1] But there was a problem there, namely that on setting out for an excursion the day before he had expected to reach this hut and was tired and disappointed that he had not reached it. This is the simplest way of solving the problem of his disappointment,

1. *The Interpretation of Dreams*, p. 108.

a way particularly indulged in by children, who have no other means of satisfying their wishes, as the adult has by his determination and will. True, the dream is wish-fulfilment, but true also that it arises from an unsolved problem. The frustration of Arctic explorers makes them dream of lavish meals, and the hopeless monotony of Negro slaves makes them dream of a heaven of eternal rest and bliss. Wish-fulfilment is one way, but a very poor way, of solving our problems and occurs very commonly when there is frustration of our impulses and desires, whether from external circumstances or from internal prohibitions. So Freud's small daughter dreamt that her little boy friend Emil was one of the family and that her mother came into the bedroom and threw a large handful of chocolate bars under the beds. In this too there was a problem which arose from frustration, for Emil was *not* a member of the family, and her mother had the day before refused to give her permission to have chocolate bars. She solved that problem by simply granting herself these wishes. Freud cannot say such dreams present no problem to be solved; they state a problem and offer a solution, though a very inadequate one.

(B) Other dreams merely revive the original experiences which caused the trouble, and go no further than a presentation of the problem: they are almost *purely reproductive without any attempt at solution*. Illustrations of this are the repetitive dreams and nightmares of childhood – of a dark figure at the foot of the bed, of wandering about lost in an empty house, of suffocation, of being chased, of separation, or of terrifying noises. These dreams may be purely repetitive as we have seen, the same dreams over and over again pointing to an early experience to which we have never been able to adjust ourselves adequately.

A young woman had a persistent nightmare of a noise getting louder and louder until it was unbearable and a crash came and she woke. Free association carried her back to a high fever she had as a child (her father reports she had a temperature of 108° F.). The noise was the throbbing of the blood in her head until the crisis came. Another had a nightmare of a convulsion in which she would in fact throw herself out of bed. This was diagnosed as epilepsy, for which she had been treated for years, but in fact it related to her having electric-shock treatment when the electrodes slipped, gave her a convulsion, and nearly killed her. The releasing of the memory of this experience in analysis cured her convulsions and her nightmare – for the function of the nightmare was to recall the cause of her neurosis, which was now achieved.

Such experiences are all too terrible for the individual to face at the time and so they are repressed. But the residual problem is still there and keeps recurring, yet only in bits; for the experiences are too terrifying to look at and recall completely – one look is enough and we wake up in terror. The function of analysis is to continue the good work and recover the whole of the experience, whereupon the patient is cured.

WHY WE WAKE FROM DREAMS

From such dreams, which merely revive the problem without being able to offer any solution, we usually wake, and this explains why we wake from some dreams and not from others. *For just as we go to sleep to forget the cares of the day, so we wake up from dreams to escape the unsoluble problems of the night.* But for that very reason these dreams from which we wake are of most value in directing us to the problem, or to the original cause of our trouble, and we constantly use them in analytic treatment for that specific purpose.

(c) Other dreams are reproductive of events or of problems (as under (B)), but they proceed to point out the *possible consequences of our action.* Instances of that are the dream of the boy with the knife, depicting the possible consequences of his rage, and the dream of the lady with the swerving car, depicting the possible consequences of her carelessness. A girl, who had a furious rage against her mother who was beating her, had a dream of a great monster coming down on her. This monster is not merely her mother; she realizes it is her own objectified and personalized rage which comes to overwhelm and destroy her. These dreams do not as yet suggest a solution, but by pointing out the dreaded consequences of our line of action they warn us against pursuing it, and act as a corrective to our conduct.

But such reminders of the consequences of our actions sometimes have an unfortunate effect, for the fear engendered in such dreams is sometimes so great that, as we have said, they end in the repression of such passions and encourage the formation of a neurosis. Biologically, they may be necessary to check our actions and emotions; psychologically, they may lead to disastrous results in producing neurotic disorders. That happened with this girl and with the boy who threw the knife.

(D) There are other dreams which appear to present a *self-portrait*, a picture of the psyche as it is; they are photographic, mirrors showing us to ourselves and in so doing helping us to face up to the problems of our psyche. Jung makes this point when, contrasting his view with Freud's, he says, 'The dream for me is in the first place the subliminal picture of the psychological condition of the individual in his waking life.' The dream of the lady in the train (p. 79) was a case in point. It showed her the real facts of the case, that she was merely bluffing herself; it made her face the fact that her husband really did not care for her, and made her call her own bluff. A corresponding dream is that of a married man whose wife and mother did not get on. There were constant rows in which he sided with his wife against his mother. But he dreams 'that he is in a car with both wife and mother and there is an accident in which his wife is badly injured: but he is solicitous about his mother and leads her away, leaving his wife to suffer.' This dream clearly reveals to him, as he recognized to his distress, that he was more attached to his mother than to his wife. It may be interpreted as a wish-fulfilment to get rid of his wife on that score. That may be so, but it was much more than that; it revealed to him the true problem which he had failed to acknowledge. The dream did not offer any solution, it merely showed the patient to himself, but in doing so cleared the way to a solution.

A woman who was devoted to her father was full of grief at his death. But then she began to have dreams of the most sordid and revolting sexual nature regarding her father, revealing an Electra complex of which she was previously unaware, but which subsequently gave her a morbid attachment to older men which was spoiling her married life. Such portrait dreams often reveal the most shocking facts about ourselves. The man who bluffs himself that he is a brave man has dreams in which he appears as a coward; the refined aesthetic dreams of filth; the cruel bully is sentimental and weeps; the good father of a family dreams of incestuous desires towards his own adolescent daughter, a dream which in fact put us on to the track of the cause of his neurosis, namely a phobia that when he takes up his carving-knife at a meal he will stab his wife. Such dreams, as we have seen, 'go by opposites', but that does not mean that they are simply the mechanical emergence of what is repressed. In emerging they reveal what is in us, the repression of which produces the neuroses. The truth is that our 'character' which we regard as normal is often a reaction character trait to opposite traits which we repress: for instance, the child who is scolded for being filthy represses this and becomes aesthetic, but it is this filthiness, which still remains an unconscious desire, which produces the obsession-compulsion to be clean and which, when his control is in abeyance, comes up in dreams in more overt form. The patient we have in mind had developed a contamination complex, washing his hands fifty times a day as a precaution against the childhood wish to play with his faeces. The dream revealed these tendencies in himself which he cannot ignore if he would discover the causes of his obsession and be restored to health. We say 'let sleeping dogs lie', and analytic treatment is sometimes criticized on the grounds that it fails to do so; but very often our dreams will now allow them to lie. Dreams bring out the most unsavoury skeletons from the cupboard of our unconscious and force us to accept ourselves as we are.

DREAMS CORRECTING FALSE ATTITUDES

There are some dreams which deal not so much with the repressed emotions, but with the false attitudes of mind which are keeping them repressed and are just as abnormal, for they perpetuate the neurosis.

A woman had collected some beautiful fresh sphagnum moss for her garden to replace some that had got mouldy. But she has a dream in which 'that new fresh moss becomes mouldy too'. Why dream of so trifling a matter? Transferring it to her psychic state, she realizes that in childhood she had adopted a mask or super-ego of being the 'nice, sweet, attractive girl', which she had assumed because it was the only way to get her father's affection– but it had long outstripped its usefulness and was a poor substitute for spontaneity, devoid of life and freshness. The dream tells her that this assumed personality, or mask, or super-ego has itself become mouldy. The dream did not solve her problem except to show her what her problem was – and to see a problem clearly is three parts of the way to solving it. It pointed to the fact that the attitude of mind she had adopted and on which she prided herself was the chief defect of her character requiring attention.

But what the dream has to tell us about ourselves is not always the 'bad' or the false; it often reveals to us the potentialities in our nature which have been repressed, but which should never have been repressed, and the repression of which has impoverished our lives. Thus the brute may dream of being sentimentally kind to a wounded animal. It is his kindly self which has been repressed. We dream of violence and aggressiveness, but if we release that aggressiveness, it turns into confidence and strength of will. We recall the dream mentioned on page 1 of the man who dreamed that he was in Trafalgar Square and all the statues came to life. This is a Pygmalion dream, and relates to those latent qualities within our nature which are at present passive, lifeless, and petrified, but which, the dream indicates, are beginning to come to life, and by the expression of which we may achieve new power.

A typical dream of this kind relating to a factual experience but carrying moral and psychological problems arising from it is the following dream of a middle-aged unmarried lady. 'I dreamt I was in a hospital as a nursing sister, having a meal with another nursing sister. I was taking my first course when she snatched away my sweet. I was furious and threw the remainder of my plate into her face. I then felt so horrified at what I had done that I went out into the yard into a donkey shed which we had as children, partly for comfort, and stayed in its cramped quarters with the donkey. The floor was covered with its manure, but in shape the manure was like meringues. Then I realized that there was somebody kind and sweet looking everywhere for me to be good to me: but I kept myself hidden so as not to be found.' This patient was the unwanted child of wealthy parents; her mother would not look at her at birth and gave her into the charge of nurses who therefore spoilt her. On one occasion a boy was chasing and tickling her older adolescent sister in the garden. The patient at breakfast brought her sister to task for this unseemly sexual behaviour, whereupon the sister said she, the patient, was jealous, which was so near the truth that the patient threw a knife at her. This corresponds with the meal with the hospital sister. She was horrified at what she had done, was condemned by everyone, father, governess, and sisters, and was filled with humiliation and self-pity. The donkey symbolized both, as a donkey is a despised animal, but this donkey was also her pet. But her association about the donkey was also that it had been castrated. This represented the fact that the experience not only humiliated her, but made her cast out all sex and love from her life, and cover herself with guilt as a protection against doing anything of the kind again – for it was the arousal of her sex feelings (young as she was, about nine) in the tickling of her sister by the boy that had caused the quarrel. The sweet thing she desired (meringue) turned into something filthy, the manure. The dream proceeded with the kindly person searching for her. This made her realize that the trouble all her life was that she had so overloaded herself with guilt that she was hiding from forgiveness and love; more than that, she would not forgive herself. All of this she herself analysed spontaneously and interpreted without any prompting by the analyst, and in fact before she related the dream to him.

In this dream there is represented the factual experience together with the psychological and the moral problems arising from it: it reveals her jealousy, her spite, her sexual desires, her humiliation, her self-pity, her remorse. It revealed also, though thinly disguised, the original cause of her neurosis, the main symptoms of which were depression, self-depreciation, and guilt with resulting lack of self-confidence. It revealed also the reason for the perpetuation of that neurosis, namely the refusal to forgive herself. It gave her a great sense of relief to realize it all, and further dreams carried on the same theme of her willingness at last to accept forgiveness and love. The dream itself, in other words, spontaneously led towards a solution of the past and present problem and with it actually brought ease of mind to the patient.

The foregoing types of dream reveal efforts made by the subconscious mind to bring to consciousness the unsettled problems harassing our personality, although we may be unconscious of them. They are all working *towards* a solution, although in many cases they have not yet succeeded. But sometimes the mere recognition of the problem at issue brings relief.

(E) *The complete successful dream* as we conceive it is one which reproduces the problems, revives their causes, and then by repeated efforts arrives at a satisfactory solution. A crude illustration of such a solution is the cliff-climber, who by repeating the experiences in imagery and dream gradually works out how best to act under these circumstances. But sometimes the best solution is for him to avoid this dangerous sport and leave it to others with cooler heads; and in such a case he continues to have such terrifying dreams of falling that he 'loses his nerve' and gives up the sport. Even such a dream serves a biologically useful purpose. But the same thing applies to the deeper subjective problems. Other dreams suggest a more definite solution, like the sodium dream (p. 82), which reassures the dreamer that if he had the guts he had at school he would be able to stand the tests of life; the dream of the woman in the train (p. 79). But perhaps the most significant and satisfactory one is the Prometheus dream (p. 85 ff.), which suggests the right attitude and the use to which we can put those basic impulses in our nature which we find it so difficult to cope with even after centuries of civilization.

In the interpretation of dreams, therefore, whilst we look to the beginning of the dream for the nature of the problem and possibly its cause, it is most important to look to the *end* of the dream, if we are to discover the true meaning of the dream, for the end of the dream suggests the solution. Indeed we often have patients who come and say that they had a disappointing or discouraging dream last night; but we discover that the reason they say so is that they are preoccupied with the beginning of the dream, which inevitably reveals the emotional and disturbing side of the dream; but when they look at the end they find it quite encouraging. Freud appears to have made the same mistake in the interpretation of the 'infanticide' and other dreams, including his own dream which we describe in Chapter 7 – for he considers the facts of the beginning of the dream (which are often a wish constituting a problem), and sometimes ignores the end of the dream which conveys the real message of the dream.

DREAMING THAT WE ARE DREAMING

Many people have dreams, but in the dream realize that they are only dreaming. What is the significance of this? Let us assume that it occurred with the lady in the train, who might have dreamt that her husband told her that he preferred the other woman, but that it was only a dream. This might be taken as a wish-fulfilment, to mean that she did not wish to accept the unpleasant truth, and so said to herself that it was only a dream. On the other hand, it might be a statement of fact and mean that she must not take the dream too seriously – that her fear that her husband preferred the other woman was only her imagination – only a dream. If one dreams of a terrifying ghost in the dark, the realization in the dream that it is only a dream may mean that we are not to take it seriously, it is not real – probably some fear of childhood. Such dreams are therefore an attempt at readjustment to the situation, an assurance that it is an irrational fear – there is nothing to be afraid of really. Just as we often wake up from a nightmare with relief to find it is 'only a dream', so in these other cases the dream itself tells us it is only a dream and not to worry.

To summarize: the function of dreams is therefore, by the mere fact of reproducing an experience and by the repetition of a problem, to stand in the place of the experience (as in the cliff-climbing); they warn us of the consequences of our action (as in the car dream and the knife-throwing); they point to the causes of our trouble (as in traumatic dreams and nightmares); they reveal the defects in our armour (as in the canoe and potato dreams); and they make us face a situation we are trying to avoid (like the lady deserted by her husband). They call our bluff, they will not let us shirk the issues, but insist on thrusting the problem before us until we attend to it, face it, and solve it. They point towards a solution of the problem (as in the sodium dream, the Prometheus dream, and the hospital sisters dream). Finally, they not only reveal but release hidden potentialities and repressed emotion in our nature, so that we are restored to wholeness. In our sleep we continue to worry out the problems of life as we do in our waking state; but dreams take up more specifically those problems which the conscious mind will not face or cannot solve, and therefore serve a biological function which no other mental process does.

THE VALUE OF DREAMS WITHOUT INTERPRETATION

But it may be asked, Can dreams be of any value unless they are interpreted? For if not, how can we say that it is their biological function to solve problems if the dreamer, as in most cases, cannot understand their meaning? The answer is that the dreamer can sense the significance of a dream intuitively although he may not be able to understand it cognitively. No doubt the interpretation of a dream adds enormously to its value, for it then enlists the cooperation of the conscious reason, and is so used in analytic treatment. But quite apart from its interpretation, a dream may have a profound effect upon the dreamer and may be of great therapeutic value.

Traumatic dreams like that of the cliff-climber obviously have an immediate effect on him, first in making him more terrified of such an undertaking and so warning him off such adventures when even the mishaps of the day would not have done so; then, as he gains experience in the dream, it gives him confidence and assurance, even without any interpretation. The dream stands in the place of experience, and that experience gives him courage. Similarly, dreams like that of the woman driving the car and the boy throwing the knife, vividly displaying the possible consequences of the action, did in fact have an immediate effect. The fact that the woman in the corridor of the train actually saw her husband go off with the other woman was in itself enough to shake her self-confidence and had a sobering effect on her pride, however much she told herself it was 'only a dream'. Indeed most of us have had the experience of being made happy or 'feeling rotten' during the day by a dream, although we not only do not interpret the dream, and indeed may not know what the dream was, or even remember that we dreamt at all.

In none of these cases was the dream specifically interpreted: they conveyed their own message; their effect was spontaneous and operated subconsciously. Dreams, like the other processes of the subconscious mind, get their effects by association, suggestion, and analogy; we 'feel' the effect upon us even without the cooperation of the reason, although, as we have said, they are the better for such cooperation. The dream of the boy who threw the knife vividly associated the idea of this aggressiveness with the horrors of a trial for murder, and that link of association of aggressiveness with horror operated even more effectively than a reasoned argument.

But these dreams relating to objective experiences are comparatively simple. What of the more symbolic dreams? These also have their effect upon us although we do not interpret them. We must remember that dreams come from the subconscious and unconscious, whose function is in the realm of feeling, and we can feel the significance and the meaning of things even if we do not understand them in cognitive terms. The very idea of symbols as in music and art is that they express something which cannot be expressed in words. The composer can convey his meaning far better in his music than if he tried to say it in words or in the form of a logical argument. So an expert musician amongst us can appreciate the meaning of a new composition when most of us do not. But let him try to 'interpret' it to us in words or in the form of a logical argument and he is at a complete loss. We must not make the mistake of thinking that we must understand something by way of conscious reasoning and explanation before it can have an effect on us. The heart, let us repeat, has its reasons of which the reason knows nothing. When in ordinary conversation we use an analogy, as we constantly do, we do not stop to analyse out or interpret what each symbol represents; the analogy and its meaning is immediately grasped. For instance, I do not need to explain that the term 'grasped', used of an idea, comes from grasping something with the hand – you immediately grasped my meaning without that interpretation, and without even recalling the origin of the symbolic term. When, in referring to a young man's getting into bad company, someone remarks, 'Birds of a feather!', we do not require the symbolism of the remark to be interpreted, but immediately understand it to imply that he was already a person of poor character and naturally attracted to others of the same ilk. It is just as if (and here also I use an analogy to make my meaning clear) – as if we are discussing with a friend how to deal with a man who has swindled us, and the friend remarks, 'First catch your hare!' We grasp his meaning without needing to interpret each item or symbol, and without asking him, 'What do you mean by that symbol?' So when we say, 'Don't put all your eggs in one basket', we surely do not need to follow this by saying, 'Eggs stand for so-and-so, the basket stands for so-and-so!' We already have the subject-matter in mind. The same applies to personal associations in dreams: when the patient dreamt of having 'sodium' injected into his foot, the word sodium was not merely the name of the mineral as it is with us, but it was *already subconsciously associated in his mind with his past experience of daring,* and so in the dream automatically aroused feelings of self-confidence, so that even in his dream he 'felt assured that he will be able to stand the test'. Thus symbols and analogies carry their own meaning to the dreamer without deliberate interpretation, although not to us without his associations.

It is precisely the same with the more complex symbolic dreams like the Prometheus dream which have less personal reference, but relate to mythical symbols. They too may carry their message irrespective of interpretation. When we in our blundering way interpret such a dream we say that the Negro represents the primitive, the female form represents intuition, and stealing fire represents rebellion, defiance. We translate symbols back to what they represent. So far, so good! But in the patient's mind it is the other way round. These feelings and problems were already present in his mind before the symbols were ever formed. The dreamer is already confronted with the problem of his primitive instincts and emotions with which he does not know how to deal; they are a threat disturbing to his peace of mind. Yet he is not clear in his mind as to what the problems are since they are of an abstract nature, still less how to solve them. Therefore the dream helps him out by saying, 'It is as if a man is stealing fire', etc. The subconscious mind, pursuing its function of sorting out and arranging the material presented to it, objectifies, personalizes these emotions and in doing so clarifies the abstract problem by putting it in concrete form. Since his subconscious mind is already representing his primitive emotions by the Negro, his intuitive feelings by the female form, when the female figure says, 'Don't kill him, but send him to a reformatory', its meaning is immediately subconsciously appreciated by the dreamer, who already has all the facts before him, although it may not be so simple a matter for us in our attempts to interpret his dream. We have to *interpret* the Negro as representing his primitive self: for the *dreamer* it already stands for his primitive self, and his intuition knows the meaning of the dream as the musician understands the new composition, though in both cases he finds it impossible and indeed unnecessary to put it into words. An analyst may have difficulty in interpreting the dream because the associations are not his; but for the dreamer the symbols are of his own making, and are already associated in his mind with certain corresponding feelings and emotions. It is just as if two people were arguing the moral and psychological problem of how to cope with our primitive instincts and emotions, and one may be in favour of crushing them down, whereas the other might say, 'After all, it is better to make use of a savage than to kill him.' The first man does not need to ask him for detailed interpretation of his cryptic remark; it carries its own meaning, and is effective in making his argument more forceful and clear. That is precisely what happens in dreams which represent the problem in pictorial and allegorical form rather than in words. The use of concrete symbols in dreams as in everyday life carries its own conviction without the necessity of interpreting the dream item by item, and can have a profound effect upon us even in the absence of such interpretation. In practice, therefore, it is quite useful, even if you cannot interpret a dream, to ruminate over it during the day, for this will have an effect in impressing upon us its emotional significance even though we cannot fully understand its meaning – just as we might a piece of music by repeating it time and again.

DREAMS AND CHILDREN'S PLAY

This method of working out problems by symbolic representation finds its counterpart in children's play, much of which is found to be concerned with working out the child's own problems by the use of concrete symbols. In their play, say of fathers and mothers, policemen, robbers, fairies, and animals, they are imagining situations and taking up attitudes (to their 'naughty dolls' for instance) without in the least knowing that they are dealing with their personal problems, that they are trying to adjust their attitudes to their own naughtiness, to the authority of their fathers, to the lack of love from their mothers, to their jealousy of baby brothers, to their own conscience. In condemning the sins of the doll the child is condemning the same sins to which she is prone, without knowing that she is doing so. Children who were bombed out during the war first suffered a period of shock and then began to play at bombing. Why? Was it a wish to be bombed? No! It was because they had been confronted with a problem created by the bombing which bewildered them and they wanted to work it out. The playing at bombing had a double effect, first by acclimatizing them to the situation and then by bringing the situation under their own control as they were able to do in their play. The working out of the problem even on a symbolic plane had the effect of settling the problem in their own minds, so that after a time their personalities were restored to equilibrium – which is precisely what happens in dreams. Moreover, when the child's problem is diagnosed, the play therapist can lead the child to a healthier solution *in the play itself* by making the characters in the play do certain things and take up certain healthier attitudes which may completely change the child's own behaviour and relations to the baby and the members of its family – and all this without the child knowing in the least that its problems are thus being worked out, without the play being interpreted, and indeed without the child realizing what its problems are. Its problems are solved symbolically in play as they are in dreams. A child's play, like dreams, reproduces unsolved problems and by that means works towards a solution of other problems.

FOLK-STORIES

The same applies to folk-stories and fairy-stories, which have a profound effect on the mind of a child and of a people without their understanding their psychological meaning; that is why a child wants to hear the same story, say, of Little Black Sambo, over and over again, for it means something to him although he may have no conception of what it does mean. The story of Cinderella appeals to the child who feels itself left out, and what child does not at some time? Jack the Giant-Killer appeals to a child's desire to be strong as a corrective to his own present helplessness. Little Red Riding Hood brings him face to face with the problem of a mother or nurse who at one time appears kindly but, like the kindly grandmother, may turn into the angry wolf at any moment. They want these stories over and over again so as to get the problems clear by imagining similar situations and so acclimatizing themselves to their situations and so work towards a solution. It is indeed because we do *not* understand our deeper emotional problems that we have to work them out by analogy, by myth, and by parable, and that is precisely the function of dreams. It is interesting to note that Christ made this distinction between the automatic effect of the parable and its conscious interpretation, for when he spoke to the people in parables, and afterwards his disciples wanted an explanation, he said, 'Unto you [disciples] it is given to know the mystery of the kingdom of God, but unto them that are without [the people] ... in parables' (Mark iv: 11). To the people he told the parable and left it to produce its own effect, for obviously he was not merely story-telling to the people. He realized that the stories would have a greater effect in the untutored mind without interpretation, for the untutored mind is the subconscious mind. But as obviously his disciples could get the most out of the parables by understanding the meaning and interpretation which he gave, as in this Parable of the Sower. Dreams are parables, symbolic stories, carrying a deeper meaning, and they have an effect on the mind even when they are not consciously understood or interpreted. The story carries its own moral.

AUTOMATIC SOLUTION OF PROBLEMS

We therefore reach this principle, that there is in the psyche an automatic movement towards readjustment, towards equilibrium, towards a restoration of the balance of our personality. This automatic adaptation of the organism is one of the main functions of the dream, as indeed it is of bodily functions and of the personality as a whole.

This idea need not cause us much concern, for this automatic, self-regulating process is a well-known phenomenon in physics and in physiology. The 'governor' in an engine is so arranged that the faster the engine goes the more is the steam cut off and vice versa, so that the engine goes at a regular speed. The thermostat on our stoves is worked on the same principle. The study of cybernetics, as this new science is called, provides us with the extremely interesting principle that not only does a cause produce an effect or result, but that the result can in turn affect the cause, by the process known as 'feed back'. But long before the thermostat was invented the body produced the same process. It is an extraordinary fact that the temperature of our bodies remains the same in spite of great variation in the outside temperature. This equilibrium is maintained by virtue of a mechanism which, when the body is overheated as in a heat-wave, throws off excessive heat by perspiration and so restores the balance. But there are innumerable instances of the same automatic regulation in our personalities which we call 'acclimatization'. Recent researches into the functions of the brain prove that the brain also has the capacity to adapt itself in this way, so that if we get a sudden shock it disturbs our nervous system and emotional life for a time and then the organism adapts itself to the situation and settles down. It is true that there are times when the shock or strain is too great and we get a 'nervous breakdown', but normally we accustom ourselves in time to new situations. We are at first shocked by social changes, by the rise in prices, by unemployment, by interference with the *status quo*, but gradually we get acclimatized to the new situation, for there is in us the mechanism which is ever at work to restore the balance.

As has been pointed out, the 'electronic brain' can be made to simulate and reproduce these functions of the human brain, not only in its capacity to solve problems, but also in its proneness to 'nervous breakdown' when it fails to do so.

The function of compensation which Jung has emphasized appears to be one of the means by which this automatic adaptation takes place, for the expression of repressed tendencies has the effect of getting rid of conflict in the personality. For the time being, it is true, the release may make the conflict more acute as the repressed emotions emerge, and we have violent dreams from which we wake with a start, and develop anxiety neurosis which fills our days with dread. But dreams are an attempt to restore the personality to wholeness and to health, contradictory as this may sound, by releasing the emotions which should never have been repressed, and the repression of which produces conflict to the personality. By this means the balance of our personality is restored.

Thus during our sleep, as the body is automatically recuperating and restoring itself from the fatigue of the day, so all through sleep the process of restoration of our emotional life is working towards a state of readjustment and equilibrium. This is mainly an unconscious process, but we are sometimes made aware of it in our dreams, and this is of great help, because being aware of the process and of the need, we may still further direct ourselves to achieve the same end. But the process of readjustment goes on quite apart from our being conscious of our dreams or from any interpretation. This may help to answer the question whether we are dreaming all through sleep. The answer may be that the process of restoration and readjustment is going on all through our sleep, but that only sometimes does it give rise to conscious dreaming, and then only in some cases is the dream state remembered on waking.

THE PROBLEMS DEALT WITH IN DREAMS

What then are the problems that beset our dreams? We may answer this simply by saying that *whatever we worry about we can dream about.*

Sometimes these worries are of a sexual nature in which our sexual desires are frustrated by the demands of social life, but more frequently they are problems, as Freud said, that arise from sexual complexes formed in early life. Because sex is the most inhibited of all our propensities it is the one which most commonly gives rise to problems and therefore one which commonly appears in dreams. But none the less, many other of our problems are concerned with the craving for power and its frustration, so that dreams often take an Adlerian form. Sometimes the dreams concern themselves with the deep-seated urges in human nature such as Jung has described in his archetypes, primordial and archaic material which we find it difficult to control or even to understand because we do not know what they are since we have never yet experienced them at all. Both Freud and Jung have based their theories on the conflict between the conscious and the unconscious although they differ as to the nature of the unconscious forces. We agree that probably most of the material which emerges in dreams is of such a nature.

But many of the problems confronting us in dreams are not unconscious at all, but are simply problems which have been consciously worrying our minds during the day which we have been unable to solve. The principle of perseveration (p. 67) is here operative. The mathematical and scientific problems, to be later referred to (p. 113), which were solved in dreams have nothing to do with the conflicts between the conscious and unconscious; and the same applies to the everyday difficulties of our work, of our social relationships, and of our moral problems. Therefore the formula that 'dreams are the language of the unconscious' must give place to a wider formula that dreams have the function of resolving our problems, whether conscious or unconscious. The dream works by sifting and sorting the problem material for presentation to the conscious mind, which then can reason things out to more logical conclusions. The outcome of this is an understanding of the problem involved and a directive as to its solution. That is the function of the subconscious mind as it is of the thalamus and mid-brain.

The dream can concern itself with *any* problem of whatever sort that besets the human mind, including the ordinary objective problems concerning our daily life and work.

Thus many of our dreams relate to the personal and domestic difficulties of the day, such as getting work, buying a house, family disputes, the care of a sick child, or problems of marriage. Dreams sometimes show a man he is really in love with a different girl from the one he reasons out would make a good wife. More subjectively, they may be concerned with the moral problems arising from conflicts in our personality and defects in our character which we consciously recognize but which create difficulties both in ourselves and in our relationships with others, and which the dream assiduously sets to work to point out, to sort out, and to try to solve. *There is therefore no problem of our lives, no worry or difficulty which may not be the subject of our dreams.* It is true that in many cases these difficulties may have arisen because of some deeper problem of our lives of which they are only instances; but in many other cases the present problem is serious enough in itself to warrant our worrying about it and therefore to dream about it. To take a simple illustration: a girl is in doubt whether to marry a young man to whom she is very physically attracted but who she feels would not give her sufficient care and affection. She dreams that 'I have an appointment to meet David, but a girl friend says I must move some heavy furniture round before I can meet him.' The latter (according to her associations) referred to her own burdensome personal problems which the dream tells her she must put straight before she can think of marrying him. The other girl is a part of herself, her intuition warning her not to be carried away by his physical attraction before she is fit to marry. Now she was quite aware of this problem, but as she was tempted to ignore the difficulties of his lack of affection because of her conscious desires aroused by his sexual attractiveness, her wiser self emerged in the dream as a corrective to her immediate impulses. Weeks later she had another dream in which her fiancé was ignoring her. In this dream she first goes off with a man with whom she is not in love and has sex relations with him. But even in the dream she feels this is unsatisfactory. She then resorts to masturbation, but again discards this as a poor substitute. Then she argues in her dreams that if she cannot marry the man she loves she will be his mistress, and in the dream this gives her satisfaction. But when she wakes she realizes that would never bring happiness because she wanted security, a family, and children. The problem is therefore still unsolved. Thus, the dream attempts one solution after another only to discard one after the other.

But she did not stop there: a few weeks later on she dreamed she was having a meal with her fiancé and a woman prepared a succulent dish for each of them. She enjoyed the dish, so did he. But there was also a chicken on the table, which her fiancé ate up without offering her any (which, she said, he *would* do!). But in the dream she tells herself she does not mind. Nevertheless, the woman who provided the meal (another part of herself) after a time exploded with rage at the way in which the girl was being treated and that put a stop to the meal. Obviously the patient had been trying to solve the problem by saying, 'Perhaps I am expecting too much, I must be satisfied with what I can get.' The dream tells her that this will not work out: she might take up that attitude, but sooner or later her feelings of resentment will burst forth and put an end to the marriage.

This dream shows her the true cause of her trouble, namely the complexes in herself which need to be removed; that sexual gratification alone will never satisfy her; that this man is much too self-centred to be a satisfactory husband to her, and that it is no use her thinking that she could put up with it. It shows how a dream can be concentrated on a present-day problem, and that a conscious one, and it shows the dream trying to work towards a solution, offering one solution after another only to find each unsatisfactory. This dream also illustrates how the dream, dealing as it does with the same basic problem, takes on new symbolic forms as new issues arise.

No doubt in most cases there are deeper problems without which the present difficulties would not have arisen. In the case just mentioned one of these problems was a childish craving for protection and security which made her reluctant to venture on the risk of marriage. But the immediate problem was concerned with her love affair as presented in her dream, which not only advised caution but showed her the necessity of removing those burdensome complexes with which her personality was furnished before she ventured on marriage.

THE BIOLOGICAL COMPARED WITH OTHER THEORIES

This description of our theory of dreams makes it clear that it is not entirely independent of earlier views; indeed it has arisen out of the earlier theories not only of Freud, Jung, and other scientific investigators, but also recognizes the truth contained in the popular theories. But our formula, we suggest, is a simple one and yet explains all the facts. Moreover, it gives to dreams a biological significance for the practical problems of everyday life, as well as for the deeper problems of life, not contained to the same degree in the other theories. It therefore agrees with these in certain respects and disagrees in others, which we may summarize.

It agrees with Freud's view of the dynamic nature of dreams, and that dreams have a meaning – although it differs as to what that meaning is. It disagrees with Freud's view that these dynamic tendencies are always sexual, and that the purpose of the dream is merely to permit us to sleep. Nor does it agree that the aim of a dream is merely the gratification of a repressed wish. In our view the function of the dream is the solution of a problem, and gives to the dream an importance for our daily lives far beyond that of permitting us to sleep. Nor does it agree that the aim of a dream is merely the gratification of a repressed wish, a wish-fulfilment. As we have maintained, the function of the dream is the solution of a problem, frequently arising out of a wish. This gives to the dream a deeper biological significance and an importance for our daily lives which it is well for us that we should heed.

Our theory agrees with that of Jung, who has stressed the purposeful and teleological function of dreams, that dreams are creative and constructive, and that there are elements in the dream which are purely archaic and not derived simply from the individual's past experience. It also agrees with Jung that dreams are concerned with the needs of life and not merely with wishes. We also agree with Jung that dreams are often compensatory, for they offer one method of solving problems by giving expression to what is repressed. But it maintains that the dream has a wider function than that of compensation, which appears to be the corner-stone of Jungian psychology, since its function is concerned with all unsolved problems, conscious and unconscious, including the practical everyday difficulties of life in many of which compensation plays no important part. Further we agree with Jung in his insistence on the purposive aspect of the dream, where the dream is leading us, but in our opinion he tends too much to ignore the causative aspects of the dream, which are usually stated at the beginning of the dream. To ignore this causative factor presented in the dream is to cripple our efforts to solve the problem, for to discover and eradicate the cause of the trouble is the surest way to cure. There is little use in discovering where the dream is leading us if our feet are to be held back by the shackles of our complexes, or buried in the clay of our past, and this indeed is a common theme in dreams.

As to the popular views (p. 7 ff.), the biological theory agrees with the reminiscence theory in that it regards dreams as reproductive; but they are not mere reproductions, for the simple reason that they change in their attempts to solve the problems involved.

It agrees to a large extent with the premonitory theory in that dreams are able to anticipate the future trends of events in our lives and show us the possible consequences of the way we are going. But what of the popular theory, in which there is a strong element of truth, that dreams arise from our physiological state or the stimulus of certain areas of the brain which contain the memories of the worries aroused in the dream? We have already stated our opinion that in itself the physiological stimulus is not sufficient to account for the specific dream, but that, say, in a state of fever when the whole brain is in a toxic state, it is natural to presume that it is those areas in which there is already tension due to pent-up emotions which are aroused into consciousness as dreams: that is to say, it is these experiences in which there is an unresolved problem. Our theory is not inconsistent with the physiological theory, but complementary to it.

But there are also those dreams which, as we have shown, arise from present objective stimuli. Are we justified in calling them problems? Surely! The smell of gas which produced the dream of the house burning (p. 8) was sufficient to produce alarm even in the subconscious mind and during sleep. The man who dreamt of himself with a paralysed arm and then woke to find himself sleeping with his arm cramped under his body had a problem caused by the pain and physical discomfort.

This also applies to the type of dream mentioned by Jung (p. 9) in which the patient dreams of some illness referring to some part of the body, like cancer of the stomach, but which is latent and as yet undiagnosed, the dream in this case being diagnostic of such unsuspected illnesses. Such strange sensations subconsciously perceived give rise to apprehension. A similar case is that mentioned under Dreams as Premonitory (p. 13).

THE URGE TO COMPLETENESS

If we accept these arguments, and they seem sound, dreams play a most important function in contributing to what is the chief end of the personality, namely its own completeness and fulfilment. It is a subject to which I called attention in the chapter on 'The Urge to Completeness', in *Psychology and Morals* (published 1920). Biologically speaking, the fundamental aim of every organism is that of its own fulfilment and completeness. If a branch of a tree is cut, new shoots spring out; if starfish are cut into two, each half will turn into a complete fish; if you injure your hand, all the forces of the blood are mobilized until that wound is healed and you are made whole. It is a law of nature. So it is psychologically: every individual has potentialities in his nature, all of which are not merely seeking their own individual ends, but each and all of which subserve the functions of the personality as a whole. Our personality as a whole, like every organism, is working towards its own fulfilment. That indeed is what we mean by mental health, which may be defined as the fulfilment of the whole personality by the release and co-ordination of all its potentialities.

But in the course of life many of those potentialities become repressed and, being repressed, emerge in the abnormal forms of the neuroses. In analytic treatment, as we shall later describe, we attempt to release these repressed emotions and direct them to the uses of life for which they were intended – and so make the personality *whole*. But dreams were attempting the same thing long before analytic treatment was thought of, and therefore dreams also, by releasing repressed experiences and emotions, are striving to solve these problems and to restore the personality to efficient functioning as a whole. However distressing many of the dreams may be, they are always working for our health. If such be the effect and the function of dreams, then we do rightly to give them a place in scientific thought and shall be well repaid for the trouble we take to understand them.

CHAPTER 5

THE SCIENTIFIC BASIS FOR THE BIOLOGICAL THEORY

Now this principle of the automatic movement towards solution of our problems in dreams may seem far-fetched to those who tend to rely on their reason alone for the solution of their difficulties, and it may seem incredible that dreams can solve problems which our conscious mind and reason find insoluble.

Let us therefore at this point correct a false impression that we may have created by our looseness of language. We say that 'dreams solve' these problems, that 'dreams tell us' this and that as though dreams themselves were a kind of personal agency working towards a solution. That is true enough on the surface, but to speak in this fashion is rhetorical: it would be truer to say that these problems are solved *in* our dreams. We must disabuse our minds of the idea of the dream as some kind of divine authority or Delphic oracle which makes infallible decisions and determines our fate. *Dreams are after all only a continuation of our psychic life during sleep, and of the problems that beset us.* What takes place in fact, as illustrated in the dreams mentioned, is that whenever we are confronted with some worry or problem in which a decision is difficult, we are in 'two minds' as to what to do. But the practical demands of life compel us to decide one way or another and we end by accepting one solution and rejecting the other which we push down, discard, and try to forget so that we may get on with the job. But even when we have decided on one course of action we may still have misgivings at the 'back of our minds' whether we have made the right decision – for there is obviously much to be said for the other side. But when we are asleep, and when the practical situation no longer claims our attention, when, in other words, we have time to reflect, these worrying thoughts return in very vivid and dramatic form, and whatever part of the mind is 'awake' during sleep occupies itself with these problems. 'Did we, after all, come to the right decision?' So in dreams the rejected side of our problem makes its voice heard and puts forward its claims; and since we are no longer under the necessity of making an immediate choice because of the practical needs of life, we may be able to arrive at a more balanced judgement.

This is clearly seen in the canoe dream after the dreamer had made a shoddy job of his painting. He might, of course, have accepted the fact and put it to rights. But he was busy and could not be bothered, so what he did in fact was deliberately to shelve the problem and say that 'it will do'. It is in such situations where the conscious mind refuses to face the problem, which, however, still worries us, that the problem returns in our dream – in this case telling him in pictorial form, if not in actual words, 'You know quite well that it is a shoddy job and that this is the way you ought to have done it.' We say 'the dream tells him', but it is after all only himself talking to himself, as he might have done (but did not) in his conscious waking life. Regarded in this way there is nothing particularly mysterious about the working of a dream: it is the voice of the discarded self. But it serves a useful biological purpose, for many of the conscious decisions of our daily lives are inadequate because they are too hasty, and give us reason to regret them. Fortunately when we do make such wrong decisions the discarded arguments come up in dreams in very vivid form, and insist on pressing forward their claims. They make us 'think twice' about it even after we have made up our minds. So the dream acts as a corrective to our hasty judgements and often induces us to reverse them by the influence they subconsciously exert.

DREAMS AND THE SUBCONSCIOUS MIND

Dreams are the manifestation in consciousness of the operations of the subconscious mind, and the subconscious mind has functions to perform beyond those, and often superior to those, of the conscious reason. This function we may call 'intuition', and it operates in our everyday life and all the time, although we may not be aware of it. We are wise if we take heed of its voice. For when we cannot make a decision about something, such as accepting a job, what to do in a domestic crisis, or buying a house, we commonly say, 'Let us sleep on it!' For we know from experience that when we do, as likely as not by the morning we find that the problem is solved and we are clear in our minds as to what to do. That is not because our brains are fresh in the morning, for the problem has already been solved in our sleep. Our mind has not been inactive during sleep but has automatically sorted out the various factors in the case and adjusted itself to the situation. That this actually happens in dreams is common knowledge: that it is *the function* of dreams to do so is our contention. Those who do crossword puzzles have often found that the right answers to many of the questions come to them not by reasoning it out, but suddenly, out of the blue.

SCIENTIFIC DISCOVERIES BY DREAMS AND INTUITION

But lest the reader should think that we are overstressing the case, ascribing to dreams so important a function as the solution of our practical, personal, and moral problems, we may draw attention to the most telling evidence, namely that many mathematical and scientific problems are solved in this way; that some of the greatest scientific discoveries have been made in dreams, and that this is a method constantly and deliberately used by scientists when confronted with a particular difficulty.

To take one of numerous instances: there is the often-told story of how the German scientist Kekulé hit on the conception of the benzene ring, an idea which has revolutionized organic chemistry. He tells us he was writing his chemical text-book. 'But it did not go well. ... I turned the chair to the fireplace and sank into a half sleep. The atoms flitted before my eyes ... wriggling and turning like snakes. And see, what was that? One of the snakes seized its own tail and the image whirled scornfully before my eyes. As though from a flash of lightning I awoke: I occupied the rest of the night in working out the consequences of the hypothesis. Let us learn to dream, gentlemen.'[1] In this case, as in dreams, that was the automatic solution of an unsolved problem by subconscious processes, and as in dreams, by the use of symbols, not by logical reason, which in fact had failed.

This is by no means an isolated case; there are many illustrations of such scientific discoveries by flashes of insight. Otto Loewi, Professor of Pharmacology, woke one night with a brilliant idea of the chemical mediation of nerve impulses. He wrote it down, but could not decipher it in the morning, to his consternation. But fortunately during the next night he awoke with the

1. W. I. B. Beveridge, *The Art of Scientific Investigation*, p. 56.

same flash of insight and this time took no chances.[1] But these flashes of insight are not always left to chance, for this method is deliberately used by some scientists, who will put themselves in a state of relaxation either with the problem in their minds or having dismissed the problem, and find that flashes of insight come to them in this way.

'In seeking original ideas', says Beveridge, 'it is sometimes useful to abandon the directed controlled thinking and allow our imagination to wander freely to day-dreams.' Cannon, the distinguished neurologist, states that from his youth he was accustomed to get assistance from sudden and unpredicted insight and that not infrequently he would go to sleep with a problem on his mind and on waking in the morning the solution was to hand. He says, 'As a matter of routine I have long trusted unconscious processes to serve me' – and would employ his sleep to solve his problems.

The mere scientific technician proceeds on the methods of observation, classification, experimentation, and generalization which we know so well from our classroom. The brilliant discoverer, on the other hand, relies on his intuition more than on his reason, and that is why he becomes the brilliant discoverer and not a humdrum technician. We might imagine that a mathematician like Einstein comes to his conclusions by hard work and logical reasoning. No doubt there is plenty of that, but Einstein himself regards intuition as being the most important factor in discovery. He says, 'There is no logical way to the discovery of these elemental laws. There is only the way of intuition.'[2] He tells us that the main generalizations of his theory came to him when he was on his sick-bed. A distinguished poet of international reputation also tells me that many of the inspirations of his poems come to him in the night, whereupon he gets up to put them down. All these illustrations prove that the mind, or the brain, has the power, apart from the reasoning processes, of automatically solving problems, especially in moments of relaxation.

That is precisely what happens in dreams: we lay aside our daily practical, moral, and other problems; the dream takes them up, and the dream solves them subconsciously more effectively than we can by conscious reasoning.

1. *The Art of Scientific Investigation*, pp. 69-70. I am informed that *The Anatomy of Inspiration*, by Harding, publ. by Heffer (Cambridge), gives many instances of this kind.

2. *Ibid.* p. 56.

How this comes about we do not know; we can only ascribe it to subconscious processes of which we know little. But recent researches in the electrical reactions of the brain confirm the fact and throw a light on it. Indeed an electronic brain can be made which can in a few minutes solve problems which it would take two or three weeks for a mathematician to solve. This instrument has extraordinary powers of adjusting itself to new conditions and of restoring itself to a state of equilibrium. This instrument, like the human brain, also gets a 'nervous breakdown' when the problems presented are too difficult to solve; for instance, when it has to take in too many circuits. In such a case, by cutting out a circuit the electronic brain may be restored to normal functioning, which is the rationale of the operation of pre-frontal leukotomy (in which a part of the brain is severed from the rest) for the treatment of obsessional states. If electric machines can be made to adjust themselves to situations and difficulties, and automatically work out problems, why not the human brain? If dreams can work out complicated mathematical problems and make important scientific discoveries, why should we think them incapable of solving the simpler problems of our personal lives?[1]

The analogy between these scientific procedures and the solution of personal problems in dreams is striking. In the first place, the mind in both cases has been preoccupied with a problem which it has not been able to solve. In the second place, this problem has usually been laid aside for a time. Thirdly, the solution of the problem comes spontaneously, out of the blue, in a period of mental relaxation, in the case of dreams during sleep. Fourthly,

1. We must not, however, press the analogy too far. Dr Grey Walter, a pioneer in electroencephalography, tells me in a private communication that his explanation of dream solution is not that these problems are worked out in the human brain as they are in the electronic brain (on which he is an expert), but that they operate by a process of association: and on reading this foregoing paragraph, one of my sons, a mathematical physicist, writes to say the analogy is not correct, for 'the electronic brain solves problems which are fed into it in accordance with definite "rules" which are also fed in'; this, he points out, is exactly analogous to the straightforward, logical, conscious mathematical methods of reason. What the electronic brain cannot do is to create ideas from an apparently formless jumble of information, as the subconscious can. The physiological analogy mentioned later (p. 117) is probably more appropriate.

when the solution has arrived we are in both cases quite unaware of the processes of mind by which the discovery has taken place. It has, in other words, come by intuition, or subconscious inference, in a spontaneous flash of insight. Finally some scientists imitate dreams in that they find it most effective to make *visual* images of their problem. Kekulé, as we have seen, got a *picture* of the benzene ring. Erhlich, whose discoveries laid the foundation of chemotherapy, made a point of making pictorial representations of his ideas. Clerk Maxwell developed the habit of making a mental picture of every problem. Thus they stole a leaf out of the book of dreams, for dreams also turn ideas into visual images, make pictures of them, and this is often found to be a more effective way of solving the problem than the method of logical reasoning.

This method is also used by the analyst in treatment, who gets his patient to lie down on the couch in a state of relaxation, for it is found that in that half-dreamy condition not only do old memories of the past and the causes of the neurosis come more readily to mind, but in that state we can see our problems more clearly and also see the solution of them. It is difficult to understand why Jung, who believes so strongly in the functions of intuition, should disapprove of this method of relaxation and free association in favour of a more fully conscious method of conversation and reasoned discussion.

We may regard these capacities to worry out problems to a solution as the work of what is commonly called 'the subconscious mind', although it would be more accurate to speak of it as the work of 'subconscious processes'.

'LAYERS' OF THE MIND: THE CONSCIOUS, SUBCONSCIOUS, AND UNCONSCIOUS

The term 'subconscious' has unfortunately fallen into disuse although it was in common circulation a generation ago. We say 'unfortunately' because the processes of the 'subconscious' differ both from that of the 'unconscious' and from that of the 'conscious' mind.

The unconscious mind as Freud and Jung have defined it consists of instinctual and other forces which have either been repressed or have never yet emerged into consciousness. The function of the subconscious mind is not merely to act as an ante-chamber in which are temporarily stored up the experiences of the day, but also to act as a selector: it is capable of taking material presented to it, whether from the unconscious mind or from the conscious perceptions of everyday life, and by so doing it works upon the problems involved, transforming and directing the energy and material thus presented to useful ends. It is not itself like the unconscious, the source of these impulses and potentialities, but it can take such energies and sift them, sort them out, harmonize, arrange, and adapt them to the everyday requirements of life.

We turn to physiology for an analogy, for there is indeed a certain correspondence (though we must not press the analogy too far) between the three 'layers' of the mind and the main structures of the brain, the fore-brain (including the cortex), the mid-brain, and the hind-brain, which have entirely different although very closely interrelated functions to perform. Each of the areas consists of distinctive types of cells, the latest to develop in evolution being the cortex, which is the operational basis of the highest mental functions of inhibition and reason. Beneath that is the mid-brain, consisting of groups of cells of which the thalamus is one. Now this, according to Le Gros Clark, Professor of Anatomy at Oxford University, 'comprises a series of relay stations through which most sensory impulses must pass before they reach the cerebral cortex.' But 'these groups of cells are more than simple relay stations; *they are sorting stations which allow for the sorting and re-sorting of the incoming impulses* so that they are projected on to the cerebral cortex in a new kind of pattern'.[1]

This is precisely what we conceive to be the functions of the subconscious mind, and it appears to be precisely what happens when a scientist has a problem concerning a mass of details; the material is sorted out (as in the case of Kekulé and the others) for the use of the conscious mind (the cortex) and thus the problem is solved. W. Penfield, the neuro-surgeon, in his article on 'The Cerebral Cortex' in the same volume (p. 63), says, 'Such a headquarters switch-board as that is so delicate, so complicated, as to stagger the imagination, but the evidence is overwhelming that it does exist.' Dreams appear to be the manifestation in consciousness during sleep of these subconscious processes; they

1. W. E. Le Gros Clark in *The Physical Basis of Mind*, p. 19.

give the answer and results of these deliberations. Dreams are not merely the language of the unconscious; they are as much a product of the subconscious mind as of the unconscious; they not only present material which has been thrown up by the unconscious, but sort out and rearrange this material and thus solve our everyday problems as effectively as they do the problem of the scientist. They not only are able to relay archaic and unconscious material on its way to consciousness; they are able to rearrange this material into ordered form for the use of the personality.

METHODS OF 'REASONING' IN THE SUBCONSCIOUS

But the subconscious mind differs in its functions also from the conscious mind, and its method of working out these problems is entirely different – that is why it is successful in solving problems where the conscious mind fails. For one thing, as we have already pointed out (p. 72), dreams have a greater effect than the events of the day because they are greatly exaggerated in over-emotional tone, and also because they dramatize the experience. For one of the characteristics of the subconscious mind, corresponding with the function of the thalamus, is that it is much more emotionally charged than the conscious mind, whose functions are more that of criticism and discrimination. Our emotions are the driving forces of our lives, and the conscious reasoning mind can no more do without the subconscious mind in its efforts to cope with life than the cortex of the brain can function adequately without the thalamus. Man cannot live by reason alone. It was, as we have already noted, a physicist and mathematician, Pascal, who said, 'Le cœur a ses raisons que la raison ne connait point'[1] (The heart has its reasons, of which reason is quite unaware).

Not only in the matter of emotional tone and dramatization, but in coming to its conclusions, the subconscious mind works on entirely different principles from those of the conscious mind. The conscious mind works by reason, by logic, by scientific deduction: the subconscious mind works by analogies, by similarities, by associations, and by suggestion. That is why when we put a person into a quiescent state as in hypnosis – that is to say, put in abeyance his conscious and critical mind, and appeal only to his 'subconscious' mind – we produce a greater effect

1. *Pensées*, iv, 277.

upon his personality by suggestion than by an appeal to reason. Again, when in analytic treatment we are trying to discover the origin of any disorder in the mind we proceed by the association of ideas – 'whatever comes into your mind' – and we deliberately exclude trying to discover what was the *likely* cause, what was reasonable, or to remember what actually happened. By the latter methods we get nowhere; by free association we can call up the spirits of the dead and so discover the real causes. The mind of the child may be observed to work by the same processes characteristic of the subconscious mind, that is to say, not by logic but by processes of analogy and similarity. Thus a child of two and a half seeing a bubble on the surface of her soup remarked, 'Look! the soup has a cold.' The child's mind does not appreciate cause and effect, but works only on the principle of antecedent and consequent, as in the case of the boy washing his hands who turned to his father and said, 'If you put your hands under the tap long enough it makes the water hot.' The child perpetually commits the fallacy of *post hoc ergo propter hoc*, because it works on this principle of contiguity, not on logic. The British Treasurer of an Asiatic community informed me that when he introduced his budget and asked his Parliament for thirteen millions, he appealed to them mainly on the ground that thirteen was a sacred number. There was no logic in his argument, but it appealed to the emotional and religious life of his hearers and more effectively secured the passing of his budget than his economics would have done.

It is this capacity of the subconscious mind to work by similarities, bringing together facts and events which apparently have no logical connexion, which leads to important discoveries, as when Newton discovered from the analogy of the falling apple the movement of the heavenly bodies in space. When we wish to clarify a difficult point we can often do so by way of an analogy. William James, the great psychologist, said that genius was mainly a capacity to see analogies.

INTUITION AND ITS FUNCTIONS

It is probably from this sphere of the subconscious also that intuition springs. We have defined intuition as subconscious inference. When a woman intuitively likes or dislikes a man she meets for the first time, she may not know why, but when it is analysed this judgement is found to be based on things that she has observed, maybe the way he smiles, or looks, or hesitates; and although these observations were not in the focus of attention, she nevertheless bases her intuitive judgement upon them. Her judgement is formed not by any conscious reasoning, but by subconscious inference: nevertheless it is based on observations subconsciously perceived. Intuition, therefore, is the capacity 'to put two and two together' subconsciously. The subconscious mind, then, like the thalamus, has the uncanny power of sorting out the material presented to it, linking up parts of it by analogy rather than by logical thinking, by association rather than by reason, by suggestion rather than by proof, and then working towards conclusions which are beyond the powers of conscious effort, because it presents a new approach to the unsolved problem.

Sometimes in fact a dream tells us to cease to make a conscious effort, to put aside reason, and to leave the problem to unconscious processes, as in the following dream: 'I dreamt I was riding a horse to a country house, but when I reached it I found that the park gates were locked and the gates were far too high to jump. I then threw the reins on the horse's neck, whereupon it leaped over the gate with the greatest of ease.' The idea of man's reason and his passions being symbolized by the rider and the horse is as old as the classics. This dream tells us that this man relied too much on his reason to get over his difficulties, and that this cramped his movements. It also tells him that if he will leave more to his intuition, to his unconscious, he will surmount difficulties that he cannot overcome by conscious effort.

Dreams are the manifestation in consciousness, during sleep, of the workings not only of the unconscious, which provides much of the material, but of the subconscious mind, and its operations proceed along the same principles. They are more than the mere reproduction of problems left during the day; they sift out the material and work out these problems by their own methods, and on principles different from those of the conscious mind; they are the products of a different method of arriving at conclusions: instead of using the process of reason for coming to a conclusion, they use the function of analogy, of simile, of parable, and of symbolism. It is for that reason, as well as the constant reiteration, exaggeration, and dramatization of the problem, that the subconscious mind is manifested in dreams, and is able to solve problems which the conscious mind by its reasoning has failed to solve.[1]

INTUITION AND DREAM SOLUTIONS NOT ALWAYS CORRECT

But it must not be assumed that the arguments and judgements of the subconscious mind arrived at by this method of analogy and suggestion are invariably correct; they often lead to erroneous conclusions just as intuition does. Nor would we suggest that the conclusions arrived at in the dream are necessarily correct ones; it stands to reason that if they are trying to work out problems dreams are always changing, trying one solution after another, as we have seen. Therefore they need thereafter to be subjected to the judgement of reason and logical proof. Einstein, as he says, could not do without intuition in his great discoveries; nevertheless, it did not prevent him from submitting them to experimentation and logical proof. To take a more humble instance, the woman in her intuitive judgement of likes and dislikes of men may prove quite wrong, because of some prejudice from a previous experience arousing emotions which do not belong to the present. Because she was once 'let down' by a dark-haired man with a moustache, she may dislike all such men although no doubt many such are quite admirable. Orators, again, particularly politicians, often misguide their audience in this way, for when they want to make a questionable point for which they have no logical argument they resort to analogy by which they hope to dupe their hearers. It is the art of the demagogue which unfortunately frequently succeeds. But this success only proves

1. The reader must here once again not be confused by the use of words; for, as we have remarked, the word 'consciousness' is used of an experience, whereas the term 'the conscious' or 'the conscious mind' is used in psychopathology as a sphere of operation. Dreams of course are not 'unconscious' – we are perfectly conscious of them – but they are derived from a sphere of mental functioning which operates unconsciously or subconsciously. They may therefore loosely be said to be the activity of the unconscious mind and the subconscious mind regarded as spheres of activity. The system of words is illogical, but no doubt the reader's intuition will enable him to understand the distinction.

our point as to the effect of such argument upon people. For in a crowd the individual's critical and reasoning judgement are in abeyance, the cortex is out of action in favour of the more emotional subcortical areas of the brain, and any argument that is brought forward appeals more to the emotions and to suggestion than to reason. On reflection many such hearers may 'feel' that there is something wrong with such an argument, but only the logician can detect where the fallacy lies.[1]

Whilst, therefore, subconscious inference may arrive by its particular modes of operation at truths which the conscious mind would not have reached, its judgements need to be thereafter submitted to the judgements of reasoning and logical proof, and that is the function of the sphere of operation we call the 'conscious mind'. Dreams should also to be similarly vetted.

Further to this, the subconscious mind labours under a grave disadvantage in its efforts to solve these problems which have been relegated to it, for the personality which finds itself unable to solve these problems and dismisses them to the subconscious or unconscious resists their return to consciousness and is reluctant to face them when reminded of them in dreams. Therefore no sooner does the subconscious and unconscious mind thrust up this unpleasant material than it automatically invokes the inhibitions of the conscious mind which finds them unacceptable or represses them again, and this does not give the subconscious mind a chance! That is why some of these experiences and problems only come up in snatches of the past where the problem originated, and also why they have to come up so frequently and repeatedly if they are to be solved. If the resistance is still stronger, it wakes us up in alarm, it will not allow us to look at the problem at all – or makes us forget all about it on awakening. But the subconscious mind has another method of making its claims without waking us up, namely by resorting to symbolism, to allegory or other distortions; this is a method by which the subconscious mind persistently thrusts forward its problems, but at the same time without evoking the hostility of the inhibiting forces. Dream distortion or symbolism, from one point of view, is a way of giving legitimate though indirect expression to repressed forbidden wishes, as Freud says; from another point of view it demonstrates the persistence of the

1. See Susan Stebbing, *Thinking to Some Purpose* (Pelican Books).

subconscious mind in presenting its problems which the conscious mind is unwilling to face even in sleep. If it cannot use direct methods, it uses indirect methods to present these problems. The use of symbolism therefore enables us to go on sleeping (as Freud says), but its main value is that by allowing us to go on sleeping it enables us to go on considering the problem when otherwise we should wake up to avoid facing it once more.

If we have given considerable space to this question, it is because it is surely time that dreams came into their own as a most important biological function in the solution of the ordinary problems of life, by sorting and sifting material by methods quite different from those of the reasoning mind. They function as a process which operates in our daily lives as intuition on which we form our judgements quite apart from reason: they are a means by which important discoveries are made in mathematics, and in artistic creation, and are used deliberately as a method of discovery by scientists and artists. If, as we have seen, the brain has the power to solve complicated problems in science and mathematics, it is not beyond the bounds of possibility that in the same way it automatically works out the ordinary problems and difficulties of life; and this in fact it does. It is a process that is going on all the time and not merely in moments of inspiration, and it goes on irrespective of any conscious effort on our part. Indeed it is when the mind ceases to make any further effort that, as in the discoveries of scientists, the subconscious mind takes over and suggests other methods of arriving at a solution. It is a process of which we are usually unaware, and therefore we speak of the results as coming 'out of the blue', like a flash. We are conscious of the problem going into the machine, we are conscious of the solution. We are quite unconscious of all the processes in between.

THE UNCONSCIOUS

We must now pass to consider what we mean by unconscious mental processes, or the unconscious mind, for dreams are said to be 'the language of the unconscious'. What is the unconscious mind, if it exists at all? The term unconscious is a term used both by Freud and Jung, and most other psychopathologists, as necessary to explain all the phenomena which emerge into conscious life. What then is 'the unconscious' or 'the unconscious mind', if it exists? Strictly speaking, the term 'the unconscious' refers in psychology to those mental processes which are devoid of consciousness – if there are such. But the term has been extended in psychopathology to signify a repository of past personal and even racial experiences, a sphere of mental functioning which is not merely devoid of consciousness, but which has its own characteristic contents and modes of functioning – the unconscious is said to consist of infantile sex wishes, of archetypes and so on. This unconscious contains material inaccessible to ordinary consciousness except by special methods such as hypnosis, free association, and interpretation of dreams and myths. This double use of the term leads to some confusion, as in the statement by Jung that the unconscious contains consciousness, and the statement by Freud that the unconscious contains only sexual wishes, but also that the super-ego or moral self has its roots in the unconscious. The problem of the unconscious is important to us because dreams are said to emerge from this sphere of the unconscious.

Psychologists long before Freud had been forced to the conclusion that there are processes within the psyche giving rise to consciousness, which are active, but of which we are quite unaware. But Freud established that these unconscious motives were the cause of the neuroses. The hysterical woman who has a pain in the back is quite unaware that this pain is motivated by self-pity, though such may be the fact; and the man with the fear of hurting may be quite unaware of the fact that his symptom represents an unconscious wish to hurt. Psychopathologists speak of this process as 'unconscious motivation', and the whole realm of thought and feeling thus repressed as collectively 'the unconscious'. According to Freud, it is the storehouse of repressed personal experiences; according to Jung, it consists also of racial experiences.

This is a convenient way of speaking clinically and we cannot very well avoid it. But are there in fact unconscious *mental* processes? When at a conference in London a clinician stated that 'the unconscious mind has all the attributes of the conscious mind, except consciousness', a philosopher retorted that there are none! – for mind and consciousness are synonymous and coterminous, and if they are not conscious they cannot be mental.

One cannot, they say, have an 'unconscious' self-pity or desire. But the facts remain that the pain serves the same purpose as the self-pity; that it is when self-pity is repressed that the pain appears; that when we trace back the pain to its source, the patient comes to recognize that it was self-pity which originated the pain although she was unaware of the fact; and finally, that when this self-pity is revived into consciousness the pain disappears. There is obviously a very close connexion between the unrecognized self-pity and the pain. These are clinical facts beyond dispute whatever the explanation. There is therefore considerable justification for the psychopathologist's speaking of repressed and unconscious self-pity, and speaking of the sphere of operations of such processes as 'the unconscious'.

But if so, how is this self-pity retained, and in what sense does it exist, when we are not aware of it? The commonly accepted idea is that these experiences have been 'stored up' in the mind, but nobody seriously believes that ideas can literally be stored up in this way – it is only a metaphor, as indeed is the phrase 'the storehouse of memory'; ideas are not 'things' that they can be stored. But if these psychic processes are not 'stored up' in the mind, and do not exist when they cease to be conscious, how are they retained? For experiences are certainly retained in some way and can be reproduced, or else we should not have memory of what we did yesterday, and we should not have a hysterical pain arising from experiences years before. In what form, then, or in what way are these experiences – whether personal, according to Freud, or racial, according to Jung – retained?

The truth appears to be that what persist are not ideas or emotions as such, but the *conditions for their revival* – just as a gramophone or pianola record makes the reproduction of a musical composition possible. When someone wants to play the pianola we say 'the music is in the cupboard' – but of course it is not the music that is in the cupboard but only a roll of paper with certain perforations. For music is only music when it is musing, that is, when it is actually functioning. What is in the cupboard is not the music but the physical conditions of its revival. Furthermore, the reproduction is not identically the same music; it is a new experience, similar to the original but never *it*. So with mental processes; they are only mental when they are actually conscious, whether thinking or feeling. But there must be certain basic conditions which retain the record of past events and which, when activated, produce ordinary memories, memories of yesterday, of childhood, and perhaps reproductions of archaic material. What then are these 'conditions of the revival' of memories, personal or racial?

THE PHYSICAL BASIS OF MEMORY AND OF THE UNCONSCIOUS

The theory of the 'storehouse' of memories gave place to the theory suggested by William James that mental experiences left behind them physiological traces in the brain which facilitate their revival. Thus a habit once acquired was easy of reproduction because the channel in the brain had already been made. Thus when we speak of our instincts of fear, aggressiveness, sex, and the maternal instinct as being 'inherited' we recognize that it is not these instincts themselves that are inherited, but only certain neuro-muscular dispositions which, when aroused by external and internal stimuli, give rise to these responses which we call the instincts. We have already referred to the glands producing the lactogenic hormone which in turn gives rise to maternal tenderness, the 'milk of human kindness'. We distinguish, therefore, between the physiological basis capable of giving rise to certain types of mental representatives and the functions themselves. We inherit our eyes; we do not inherit our sight. We inherit only the conditions for the production of hearing, not the hearing itself. We do not inherit our digestion, we only inherit certain nerves, muscles, and glands which when stimulated into activity produce secretions and digest our food. We inherit the physiological condition of fear, not fear itself. As a general principle, however, we may say that in so far as our mental processes depend upon our physiological processes they may for practical purposes be regarded as inherited. But this physiological theory may now have to give place to a suggestion from physics, which simplifies the problem and gives explanations for things otherwise unexplained. I have seen suggested somewhere the hypothesis that when we have an experience it sets up in our brain a *circuit of electrons*, which continues to operate even when consciousness of the event has completely disappeared. The idea of these circuits of electrons in fact corresponds very closely with our experience; for when we have a worry we find that the idea 'goes round and round in our heads' – although that is probably only an analogy.

This circuit of electrons is continually active. At the same time it is itself unconscious in the strict and literal sense of the word, devoid of consciousness, but can, under appropriate conditions be revived to give rise to conscious processes reproducing the original experience or something very like it: and this is what we call 'memory'. This circuit of electrons is therefore not itself a 'mental' process; it is unconscious in the true sense that it is not conscious, being a physical process, but, like the pianola or gramophone record, is a permanent and necessary condition of the revival of our experiences. This concept is much more satisfying than the idea of ideas being stored up, or even the idea of channels in the brain. It should satisfy the philosopher who holds that what is not conscious cannot be mental, for it does not involve us in assuming that there are mental processes which are not conscious, for these circuits of electrons are in the true sense unconscious, and are not themselves mental but physical processes. At the same time these circuits of electrons are continually active, which corresponds with what the clinicians have long maintained, that complexes (the psychological counterpart of the circuit) though repressed and forgotten are still active; that they affect our conscious life although we are unaware of them, and indeed that they can produce neurotic disorders. Thus unconscious motivation becomes a physical reality.

Further, physicists tell us that such a circuit of electrons may be interrupted by a spoke in the wheel, as it were, in which case the electrons are dissipated. This corresponds to a well-known clinical experience that when we revive some old experience, such as a fear in childhood or a later dramatic experience like 'shell-shock', which originated a neurotic symptom (like claustrophobia, or the self-pity underlying a hysterical pain), the symptom disappears, a fact to which Freud gave the name of 'abreaction' – letting off the emotion. The bringing up into consciousness of the forgotten experience tends to dissipate it. But the whole idea, attractive as it is, is purely hypothetical.

Granted the existence of these circuits of electrons as the physical basis of memory, we may further conceive that while some of these circuits are the result of our individual experiences, others are established in our brains as the result of racial experience, which may be inherited in the same way as the heart, lungs, or liver function in accordance with our constitutional make-up. Of such may be what we call the instincts – and other archaic behaviour. They may even be conceived as capable of producing images – such as the archetypal images.

It is possible, therefore, that these undischarged circuits of electrons have effects upon our daily life without our knowing it: they may emerge as memory of individual events; they may also emerge as dreams whether of individual or racial experiences; or give rise to myths; or they may without our knowing it produce neurotic symptoms.

But the fact that these processes are basically physical does not compel us to give up the problem-solving function of dreams, for it has been proved that the body itself has the power of *automatically* adjusting itself and the whole personality to new situations, by the process of 'feed-back' (p. 102) or what we more commonly call acclimatization or 'accustoming ourselves' to things.

CHAPTER 6

THE STRUCTURE OF THE DREAM

FROM the considerations described in the last two chapters we may analyse the structure or composition of the complete dream. The first part gives the problem, or reproduces vividly the experience which precipitated the problem, or it may revive the neurotic symptom, such as a fear, physical pain, guilt, or sex perversion from which we are suffering. The middle part of the dream makes tentative efforts to work out that problem and wakes us up if it cannot do so. In the last part of the dream an attempt is made towards a solution, sometimes abortive, sometimes illusory, sometimes working to a complete and adequate solution. A reference to dreams previously mentioned will illustrate this point. Patients, as we have said, only complain that they have had a discouraging dream, but that may be because they are paying attention only to the first part of the dream which contains the problem and which is therefore indeed vivid and distressing, and ignoring the latter part of the dream which is showing the trend of the subconscious mind towards a solution.

SERIES OF DREAMS

Sometimes the problem and its suggested solution are contained in one and the same dream, but frequently the problem is stated in one dream and the attempts at solution in subsequent dreams of the same night or subsequent nights. We usually find, therefore, as Freud points out, that the dreams of a night are connected in whatever different forms they are presented. As an instance: a man dreams that he goes into a very dark valley with lowering threatening clouds and deep depression. Following that in the same night he has a dream in which he is in the army mess and there is a dispute about a table; he wants to have a table to himself, and the other officers refuse to let him have it. The last dream of the same night is that he was fighting a friend of his with whom he has been friendly for fifteen years and he knocked out this man. These dreams have no apparent connexion till we discover their associations and see their meaning.

The depression, the valley with threatening clouds, was his neurosis, for he suffered from depression. That is what the dream is about, the trouble in the psyche which the dream is going to deal with. What then is the cause of this trouble? The second dream answers this by pointing to a defect in his character of which he was unaware: he must have a table to himself, in other words, he is selfish, self-willed and anti-social; that is why all the others are against him, which is the immediate cause of his depression. And its solution? He knocks out his friend. The patient's immediate association about this friend was that he was 'too big for his boots' and the patient and another colleague were in fact considering how to approach him and bring this home to him. But why was the patient so concerned about this other man's arrogance, and why should he dream about it? Because the other man stood for himself: for he, unknown to himself, had the same arrogant tendencies as he had in fact been a spoilt child. The first part of the dream pointed this out in making him demand a table for himself in the officers' mess: he also was 'too big for his boots'. The dream brings home to him the fact (*a*) that he was selfish and arrogant, (*b*) that arrogant people like him and his friend are not tolerated, and (*c*) he then realized that it was because of his arrogance that he was unpopular. Finally, (*d*) the dream points to the remedy; it makes him get rid of his arrogance, as symbolized by his knocking out his arrogant friend, for only by so doing can he get on better with his fellows and rid himself of his depression.

In such an apparently inconsequential series of dreams we see other characteristics of a dream, how a dream shows a man to himself and reveals his true nature (dream portraiture, p. 91): we see how the persons of a dream, like the selfish friend, may stand for some quality or characteristic of the dreamer himself (a point we shall take up later, p. 147 f.): we see how the dreams of a night represent a series, the first one giving, as in this case, the nature of the problem, then its cause, and finally working out to a solution; how a dream deals with a specific instance as an illustration of a whole principle, pointing to a defect in character; and the teleological or purposive nature of a dream.

This dream also illustrates the difference between the latent and manifest content of the dream: the dream arose immediately from the fact that he was planning with the other man how to bring home his arrogance to his friend. But in doing this his own conscience was awakened about himself. When he was criticizing his friend, something within him said, 'But what about yourself?' The manifest content was about his friend, the latent content about his own problems which had been precipitated by his concern about the other man's faults. The dream apparently dealt with an objective situation; in reality it was about a subjective problem of his own selfish demands, which the dream says he must get rid of. If, therefore, we have a series of dreams in the night, we find as a rule that however different in form and symbolism they may be, they bear upon the same theme, and the same may apply to a series of dreams extending over months or even years, for if we cannot find a satisfactory solution to a problem, we go on dreaming about it; hence repetitive dreams.

WAKING FROM A DREAM

Frequently, however, the dream states the difficulty and tries to solve it, but has no solution to offer, *in which case the patient wakes up from the dream.* For, as we have said, just as we go to sleep at night to escape the worries of the day, so we wake up to escape the unsolved problems of the night. This applies to both objective traumatic dreams like that of a car accident, and to subjective dreams which arise from the fear of our forbidden impulses. Therefore when we have a dream from which the patient wakes up it is an indication that the dream can offer no solution. But such dreams, which include nightmares, are particularly valuable as indicating what are the problems which are troubling the dreamer and which he is being called upon to face, and we can make particular use of them for that purpose.

PROGRESSION IN DREAMS

The solution which a dream offers is not necessarily satisfactory any more than are the intuitive 'brain waves' of the scientist, which more often than not come to nothing. It may only produce a restless night. In such a case the subconscious mind tries various solutions till a satisfactory one is reached. That is why the same theme keeps on recurring in different forms: as varied solutions are offered, the dreams take different shapes. If therefore we keep a record of the patient's dreams in analysis over a period of months or years, we find the same theme running through, but there is a progression throughout the series, so that the dream changes not only in form but also in attitude of mind and approach to the problem. Then, as each problem becomes solved in analysis, there is usually a lull in which there is an absence of dreams and then the theme itself of the dream changes as a new problem arises. Anxiety dreams change into aggressive dreams, dreams of self-pity into sympathy for others.

A beautiful illustration of progression in a series of dreams is the following: A woman dreamt that while she was having analytic treatment, there was a tiger loose in the adjoining room which threatened to break down the barrier and endanger her life. (The tiger was the repressed sex feelings which she feared.) In a dream some nights later she was in a swimming-bath and in the water was also a dog, which was repugnant to her. (Her sex feelings are no longer dangerous but still unpleasant.) Then she dreamt of cycling along a country road and lying in the road was a cow, so that she had to alight and walk round. (Her sex was now neither dangerous nor unpleasant but merely obstructive.) Weeks later, she dreamt she was awakened at night by what she took to be burglars in the garden and she called out in alarm to her sister. Whereupon the men laughed and said, 'We have not come to harm you! We have taken your luggage to the station and have come to be paid.' This indicates that her sex instincts were now to be regarded as neither dangerous, repugnant, nor obstructive but could be of definite service to her – in fact she owed them a good deal. The change in animal symbol represented a change in her attitude towards sex, to the animal part in her, and a gradual solution of her problem, culminating in her acceptance of her sexuality as making an important positive contribution to her life and health.

Another patient dreamt at one time of trying to catch a train and always missing it, which signified that he had difficulty in keeping up with life; but later he dreamt that he just caught it which implied that he was progressing. Finally, he dreamt of driving the train, that is, managing his own life successfully. Here the symbolism of the dream is the same, but the theme changes as he gets better. A male patient will have sexual dreams which are at first masochistic, of being beaten; later they are sadistic, of beating others. This is not a final solution, but it shows progress towards normality. Finally, he makes love in normal fashion. In analytic treatment we may therefore gauge the progress of the patient by the progressive change in his dreams, which tells us how the patient is getting on.

Another series of dreams illustrates this progression. A man suffering from blushing due to feelings of shame and inferiority derived from exhibitionistic indiscretions in adolescence dreams as follows: 'I am among a large concourse of people, and one man there begins to take off his clothes till he is naked. No one seems to notice it or bother about it; but I am shocked, and become a policeman and arrest him.' This first dream of the series presents the problem of his repressed exhibitionism and makes an abortive attempt to get him to take a more tolerant attitude towards it ('no one notices it') but fails to do so, for the dream makes it clear that the patient is identifying himself still with his censorious moral super-ego (he becomes a policeman), and still condemns this perverse tendency. In the next few analyses we proceeded to deal with the causes of his shame and fear of discovery in adolescence; and a few nights later he dreamt: 'I was accompanying, with other people, two men, one black and one white, who were going to fight on the seashore. They both started to strip naked. I did not mind in the least, but said, "I had rather it was them than me." ' The black man and the white represent his 'bad' and 'good' selves: but now he is indifferent regarding their exhibitionism – both the good and the bad strip, but he is now not taking sides and keeps detached from them, refusing to identify himself with either of them: 'rather them than me'. The dream goes on: 'They had their fight and went off, and then I noticed that there was a policeman lying on the sand complaining bitterly that the tide had come up and soaked him' – a ludicrous picture but an important change, for the policeman is no longer identified with himself as in the first dream; that is, he no longer identifies himself with the censorious side of himself, which, instead of being the dominant authority, is now itself in a bad way, smothered by the tide and an object of ridicule. The patient is obviously getting rid of his self-righteous attitude: 'I ignored him but was amused at this and went off' – a great change from the first dream. The patient, now identified with his natural self, leaves his over-moral super-ego to its fate on the shore. It may be objected that it is surely wrong to get rid of moral values.

The dream was quite right to bring to nought his censorious moral self, for his morality was in fact entirely self-righteousness, based on narcissism, as he was admired for being a good little boy but was in fact a little prig. It may also be objected that to be an exhibitionist cannot be right, to which the reply is that such neurotic symptoms are always the result of repressing natural tendencies which are thereby perverted. It is only by releasing them and accepting them that they can be developed and sublimated to higher uses.

The later progress of the patient is noted in a dream of several weeks later. 'I am leaving home [leaving his past behind] and on the road meet a party of students with their professor [the patient had wished for an academic career, but this had been denied him]. I joined them and was happy [a wish-fulfilment]. Then a ragged tramp joined us and I was afraid the others would think we were together and that I also was a tramp. [He is still here afraid of his past turning up and disgracing him; he is still maintaining his snobbishness though not his self-righteousness.] But I soon discovered that the tramp turned out to be a very decent, capable, and clever fellow.' The tramp of course represents his more primitive natural self which he had repressed, the outcast of which he was ashamed and the subconscious awareness of which produced his blushing and his feeling of inferiority. In the dream he is brought to accept this crude but natural part of himself and he is assured that he need no longer regard himself as an outcast, but by accepting this part of himself may become a capable and efficient person, for only by the use and direction of our primitive impulses can we be really efficient. This is the same theme as that of the Tiger dream (p. 132) and of the Prometheus dream (p. 85 f.), for the problem of coordinating our primitive impulses with our social and moral sense is perhaps the most fundamental problem in human nature. By accepting the tramp in himself he did in fact lose his shame and his blushing, and did become 'a decent, capable, and clever fellow', for he succeeded after the age of thirty in taking a university course as he had always desired. His progress in this series of dreams, spread over several weeks, clearly demonstrated how dreams start off with the problem, showing clearly what the problem is, and then help towards a working out of the problem towards a healthy solution. As the analysis helped the dreams, so the dreams helped the analysis.

THE DREAM ANTICIPATES

The fact that we dream of a certain attitude, like this man's attitude towards the policeman on the shore (his censorious self), or towards the tramp (his primitive natural self), does not mean that we have already reached that stage of readjustment in conscious life, otherwise we should not dream it. But it represents what is going on in the subconscious mind and what is about to emerge, so that what we dreamt about last night may come to pass the day after tomorrow. Such dreams, however, are satisfactory because they show the trend of coming events. They are in this sense premonitory, the promise of better things to come.

SYMBOLISM IN DREAMS AND ITS PURPOSE

Very rarely does a dream say what it means. It prevaricates, it hedges, it casts innuendoes, it hints unpleasant things about us without saying so, it slyly makes remarks about other people when it really means us. It does not speak outright, it talks in symbols, it speaks in parables, it uses analogies. If the dream wants to say something, why does it not say so plainly and not go all this roundabout way only to confuse us as to the meaning of the dream and make interpretation of the dream very difficult? The symbolic presentation of the dream is indeed the main trouble about dream interpretation and, as we have seen, largely gives rise to the difference concerning dreams between the various schools.

Freud, some of whose views on symbolism we have already discussed (p. 20), maintains that the purpose of the symbolism in dreams is to disguise their real meaning and to make the forbidden wishes of the unconscious more presentable to consciousness – so allowing the patient to sleep. It is a way of evading the censor and gratifying the wish without appearing to do so. That is obvious in the girl's dream of the bayonet (p. 18). Jung, on the other hand, maintains the opposite; that the function of dream symbolism is to bring the facts of the unconscious more vividly to mind, to impress more strongly upon us what the unconscious is trying to say. Symbolism, he says, is the 'process of comprehension by means of analogy' ... 'the symbol in the dream is approximately equivalent to a parable; it does not conceal but it teaches.'[1]

1. *Analytical Psychology*, p. 309.

Ordinary symbols in fact perform both these functions: at one time they are used to clarify our meaning, at another to disguise it. A girl's blushes both reveal and are the result of an attempt to hide her feelings. A symbol is a physical representation of an abstract idea or an emotion which we may find it difficult to express otherwise. Concrete or pictorial symbols are often far more expressive than word symbols, and an analogy often clears our mind for the understanding of our abstruse subject. A flag may arouse feelings of patriotism much more effectively than the word 'patriotism'. The analogy of a deep lake may suggest the idea of the unconscious mind much more comprehensively than the word 'unconscious', bringing before our minds, as it does, the picture of vast depths, and of living processes going on in these depths which are not visible on the surface of consciousness and of which we know little, but which every now and then cause trouble on the face of the waters. The first function of symbolic language is therefore to make clear such ideas and to express feelings and emotions which cannot be adequately understood otherwise. So we kneel to symbolize reverence or respect and the act of kneeling encourages the reverent spirit – that indeed is its intention. The act is more expressive than words.

But symbols may have precisely the opposite effect, as Freud says, namely to disguise the meaning. Thus we may kneel to pretend to a reverence we do not possess, to hide our irreverence; we are ingratiating towards a man on whom we intend to play a dirty trick; we kiss the cheek of him we are about to betray. The symbol, instead of being an aid to thought, is intended to disguise our thought; instead of encouraging an emotion or feeling, it is adopted as a substitute for it. Thus one Russian may have his walls plastered with pictures of Stalin as expressive of his sincere admiration and to inspire his loyalty, whilst another covers his walls with the portrait as a pretence of loyalty and to disguise his feelings of disloyalty.

The very nature of symbols, and perhaps one of their attractions, is that they often have a double meaning. The symbol of a lake is regarded by some as always referring to uterine waters, by others always to mean the unconscious. But water may represent *both* ideas, for it may be chosen as an analogy of the unconscious because of its hidden depths, but also because it suggests the uterine waters which, *like* the unconscious, are the source of all our creative power, the womb from which springs all life.

There are, however, certain symbols, like a tree for life, a train journey to represent progress in life, and the snake for the male sex organ, that are so appropriate that they are very commonly seized upon by the psyche to express such ideas. Such symbols are fairly universal and may justify an arbitrary interpretation, as long as we keep in mind the possibility that they may refer to something specific in the particular individual.

THE NATURE OF SYMBOL

The idea of dream symbolism is not too difficult to understand if we realize that many words and phrases were themselves originally taken from material objects and actions, as we have already pointed out. We get 'into contact' with people, we cannot 'swallow' certain suggestions, we 'rise up' in wrath, we 'cast care to the winds'. There is hardly a sentence which does not express itself in such obvious symbols – as in this very sentence, for the word 'hardly' refers to something hard as meaning something difficult, the word 'express', to push out, and the word 'obvious' suggests something staring us in the face. Such phrases represent abstract ideas, but they are obviously taken from material objects and actions. All that happens, therefore, when in dreams we wish to express such abstract ideas, is that we revert to the material object or action from which the word expressing the abstract idea was originally derived. We say of a man's reprehensible conduct, 'He makes me sick', or describe his behaviour as 'nauseating', in which we are using symbolic language taken from our digestive reactions, since we reject from the stomach what is unpleasant. Dreams reverse the process and translate the idea into the original event: so when – in a dream – we wish to express the fact that we find a man's conduct unpleasant and nauseating, we dream (as in one case) of visiting his house and as we leave we vomit on the doorstep. Furthermore, dreams set out to express ideas, not in words, which they rarely use, but in visual form, in pictures. The word 'progress', an abstract term applied to civilization or to thought, as in 'the progress of science', simply refers to an object 'moving forward'. What then is more natural, if we wish to express in a dream the idea of 'progress', than to give expression to it in the form of a train journey; and if the dream wants to express the idea that we are not progressing in life psychologically as we should because of something within us arresting our development or holding us back, it simply represents us as failing to catch the train because our legs are rooted to the spot. There are numerous illustrations of this process by which we first apply to abstract ideas phrases which we have originally taken from material experiences and then in dreams transfer them back to the original physical form when we want to express those abstract ideas. Thus we speak of a man bristling with rage, since an animal's bristles stand straight when it is aroused; that is to say, we call an emotion by the name of one of its manifestations. In the dream the anger is expressed by the actual bristling. So in dreams we actually explore avenues, we turn over every stone to 'leave no stone unturned', and literally 'walk into danger', such as over a precipice, although the danger may be a moral danger: we throw care (in the form of baubles) to the winds, and when we feel overwhelmed by an emotion, the dream depicts us as actually overwhelmed by a wave. A clown is a man whose stupid antics make us laugh: therefore if the dream wants to tell us that we are being silly, it depicts us as a clown; our self-righteousness is represented as a policeman, our slyness as a spiv. Thus symbols, by expressing themselves in such concrete and realistic forms, can be far more effective than words, and can have a correspondingly greater influence upon our lives because of their more primitive mode of expression.

This tendency to express thoughts by concrete symbols is what the orator makes use of and it often leads him to mix his metaphors, for he makes use of a different concrete symbol for each new idea. Such translation of verbal thought into pictures leads to such absurdities as saying. 'He can't open his mouth without putting his foot into it.' Dreams too are always mixing their metaphors, and this is what makes dreams so bizarre.

Let us then concoct a dream out of an idea, and for this purpose let us take the memorable instance of the speech of the member of Parliament: 'I smell a rat, I see it floating in the air, but mark my words, I'll nip it in the bud!' The speaker might have put the same idea into more abstract terms by saying that 'false rumours are apt to spread their evil influence and must be checked at their source' (even so he would be using physical analogies), but that would not have roused his audience to the heights of enthusiasm to which no doubt they were aroused by his vivid metaphor. When he put his ideas into such concrete language, these flights of metaphoric fantasy were far more striking to his audience and incidentally endowed him with immortal fame. But let us suppose that a dreamer wished to express the same idea. He would go a step further than even this metaphor and might present it in this pictorial form: 'I dream of a number of people; they seem to be looking for a bad smell, a dead rat or something. Then the atmosphere becomes very sinister as though of something dangerous floating about which we are trying to avoid like some plague; then the scene changes and we find ourselves in the garden vigorously nipping buds – a silly thing to be doing, and I don't see how you can make any sense out of that anyway!' These three pictures are apparently unconnected and quite ridiculous until we have the key to decipher the meaning, which we might get from the patient's free association, and then we discover that he himself (or his political party) was subject to some rumours which he was inclined to ignore. But his subconscious mind refuses to allow him to ignore them and urges upon him the necessity to put a stop to them immediately.

Thus the subconscious and unconscious are always trying to force into consciousness problems which we cannot solve or which we are for other reasons trying to evade, but as they cannot do so in speech they do so in pictorial and concrete language, presented in symbols, which have the effect of thrusting up the material more vividly and so more effectively. But since the mind is unwilling to look at these problems, the symbol is a means of expressing, while at the same time disguising, the true meaning which the dream is trying to express. The dream, by the very fact of using symbols, is finding a means of expressing what would otherwise be denied expression. Would we then say that the function of the dream is to express something or to hide something? It is both at once – for by disguising it, it is able to keep an unpalatable problem to the forefront of our mind.

DREAMS AND THE PRIMITIVE MIND

But the most obvious reason for our dreaming in symbols is that dreams are a form of primitive thinking. 'Symbolic language', says Jung, 'is primitive language.'[1] They arise from a primitive part of our mind which, like the mind of a small child, can think only in terms of concrete symbols and not of abstract thoughts. This is understandable from the physiological point of view, for in sleep the higher and latest developed part of the brain (the cortical area) goes to sleep first, and any mental activity is carried on in the lower (thalamic) areas, primarily the seat of the feelings and emotions. The still more primitive centres which have to do with the vegetative functions like breathing, sweating, and circulation remain active, though subdued, all through normal sleep. But in nightmares even these vegetative functions are profoundly disturbed by our emotions and we sweat, palpitate, and suffocate. The result is that any thinking which takes place in sleep follows primitive patterns of thought.

Now the main characteristics of primitive thinking, as of the lower centres of the brain, are that it is strongly emotional; that it expresses itself in concrete symbols rather than in words; that it expresses itself in sensations and perceptions, especially visual sensations, rather than in ideas, and that it is not logical but follows an associational rather than a logical order of events. It therefore cannot express such ideas as cause and effect, nor can it appreciate the essential relationships between things, but accepts external similarities as representing real relationships. Dreams, arising as they do from the lower and more primitive layers of the mind, are subject to all these strictures.

(*a*) Primitive thinking is not abstract but concrete, and that is one reason why dreams take the form of symbols, for symbols are the material representation of feelings or ideas that we cannot, or which we are forbidden to, express otherwise.

(*b*) Again, primitive thought takes place much more on the plane of sensations and perceptions than of ideas. Its knowledge of the world comes through touch, smell, taste, and especially sight. Therefore when the dream has to express these ideas in primitive language it has to put them into one or other of these forms of sensation. Most people are visualizers, since vision more than any other sense in primitive life brings us into touch with external objects. It is therefore the principal organ of perception; so much so that we say 'I see!' when we mean 'I understand'. Most dreams therefore take a visual form, and a good deal

1. *Analytical Psychology*, p. 311.

of 'dream work', as Freud calls it, consists in putting our ideas and problems into visual form, and in dream interpretation we have to transform them back. People born blind have never visualized and therefore cannot, of course, have visual dreams; so they dream mostly in terms of touch, as I have found on inquiry, for touch is the sense by means of which they have most 'contact' with the material world. A few people dream not in visions but in words. One patient dreamt simply in sentences: 'The fever is pleasant', 'The nature of the universe is subtle', and we then had to work out what his subconscious mind meant by these phrases. In others, smell plays an important part, while in others, hearing. A patient was obsessed in her dreams with the smell of a corpse which she could not locate until she discovered it in a childhood experience. Freud's original analytic case suffered from the smell of burnt pastry.[1]

(*c*) Another characteristic feature of primitive thinking is the lack of ability to relate cause and effect. In rational thinking we say, 'This happens because that happened', and we see a causal and necessary relationship between the two. The primitive mind cannot argue in that way, for the idea of cause and effect is an abstract idea and primitive man (like the child), being unable to appreciate or grasp the deeper meaning of causality (not that anyone really can), simply assumes that because one thing happens *after* another, that is caused by it. If he breaks a taboo and a storm breaks loose, the storm occurs because he broke the taboo. To him antecedent and consequent mean cause and effect. If the savage is fishing in grounds which are forbidden him and a shark comes and destroys his canoe, he considers that it is *because* he fished on forbidden ground that the gods were angry with him. Even civilized people argue so, but it sometimes leads to nonsensical conclusions. Most people die in their beds, so we should avoid going to bed at all costs!

As we have observed in the previous chapter (p. 119), we constantly find children arguing in this pre-logical way and linking things causally because they are linked in time. Because a child got angry with the mother and then the mother died, he thought his anger had killed his mother, and this had, in fact, a profound and disastrous effect upon his life, for he had killed the one thing he loved and on whom he depended. Being innocent of the laws

1. *Papers on Hysteria*, p. 15.

of logic, the child constantly commits the fallacy of *post hoc ergo propter hoc* (after this, therefore because of this). This is also notoriously true of orators and demagogues, in whose appeal to emotion there is often a complete triumph of prejudice. During the war I heard one politician before an enthusiastic audience say that 'We must bomb the German cities out of existence and that will prove who is right!' It does not, of course, prove anything of the sort.

The complete absence of the laws of logic from his reasoning makes the mental processes of primitive man difficult to understand. One instance, already given, is that of the Red Indian (p. 3) who insisted that the missionary had taken his pumpkins: he had dreamt it, therefore it must be so – that should be clear to anybody. The missionary, however, would persist in his queer white-man's way of arguing that as the pumpkins were still there they could not have been stolen! – that may be logic but did not make sense to the Indian.

Dreams, coming as they do from primitive sources, argue in such quite illogical ways. They have their own laws of argument and we cannot expect to understand their meaning until we understand their language, otherwise dreams just make nonsense as they do to most people. It takes all the skill of the anthropologist to understand primitive man's way of thinking and so to talk in his language, and it takes all the skill of the psychologist to understand the interpretation of dreams – and in neither case has complete comprehension been achieved.

Imagine yourself trying to carry on an argument with someone who has no sense of reason but who can only draw pictures, and you will realize how difficult it is for dreams to express what they are trying to say and to make themselves understood. They have to resort to ways of thinking and arguing quite different from those of logical thinking.

Dreams, like primitive thinking and children's thinking, are illogical, and therefore when a dream wishes to convey the idea of cause and effect it can only do so in terms of antecedent and consequent, by putting one thing before the other. Therefore in interpreting dreams, as Freud has shown, when one thing precedes another in the dream we may take it that the one is supposed to be the cause of the other. Instances of this are to be found in the dream of the officer at the beginning of the chapter, in which a sequence of events related to antecedent and consequence, to cause and effect. His subconscious mind wanted to impress on him that his depression was the result of his unpopularity due to his arrogance, which he must get rid of: it does so by first depicting his depression (in the picture of the valley), then his arrogance (in his demanding a table), then his unpopularity (in the argument), and finally knocking out his arrogance (in the person of his friend).

Another instance: A patient sees herself in a hard stiff dress in which she can scarcely move; after that she sees herself in hospital. Then she is in the Lake District climbing a mountain which she finds difficult; and her father is dead. This dream as it stands is rubbish, but remembering that the primitive mind works in symbols to express abstract ideas, and that cause and effect are represented by antecedent and consequent – and knowing something of the patient's life and problems – we can recognize that her dream is trying to say pictorially what the logical mind would express in words by saying that her life has hitherto been too restricted, too cramped and hide-bound, so that she 'can scarcely move'. She has lived in a strait-jacket. The consequence of this is that she develops a neurotic illness (the hospital). That is the problem of her dream and of her life. The dream goes on to say that by considerable sacrifice and ceaseless effort (climbing the mountain) she can be free, and then her father's tyrannical influence, which was the direct cause of the illness, will die.

(*d*) This leads us to another mode of thinking which primitive man shares with the child, namely that he argues by analogies, often by quite superficial analogies, and by such means arrives at quite illogical but to him convincing conclusions. Things that are similar to the same thing are similar to one another. If you want to kill a man, you spear him. But if you do not get the chance, you make a clay image of him and stick things into that, and that is just as good. The clay image is *like* the man, and therefore it *is* the man: if, therefore, you spear the image, he *is* dead. So a child being annoyed with her unreasonable father will put up a cushion on the sofa and knock it over – with marked satisfaction. Dreams in the same way, being a primitive form of thought, bring things that are similar together and in doing so many conclusions are drawn, some of them nonsense, but many of them of real value to life. Another way in which dreams impress things upon us is to bring together scraps of similar experiences from the past which apparently have no logical connexion, but when we investigate them we find that they all contain some common factor. So we may dream of an event at the age of three taking place in a house which we did not live in till we were ten, which is intended to link together two experiences of the same nature – a case of condensation. So, when we have been impetuous during the day, dreams may recall many similar experiences of the past in which we made mistakes because of our impetuosity, the cumulative effect of which compels us to face the situation. The dream does not *reason* that your impetuosity will lead you into trouble; it merely recalls the many similar instances of the same things having occurred – and the result is even more effective than merely putting one thing after another, and even than reasoning would have been, because we are presented with concrete instances of what the dream is trying to impress upon us. An excellent illustration of this is in Freud's dream which we discuss in the next chapter: it tries to convince him of the present failure by recalling mistakes of the past.

This calling up ideas by similars accounts also for the numerous puns which appear in dreams, as already illustrated (p. 81 ff.). A pun depends on a similarity of sound but a difference in meaning. When a dream wishes to imply that the dreamer is hiding something, it may depict him as going into Hyde Park; when it makes him stand boldly before the world, it makes him wear a shirt with a 'stiff front'; when it wants to refer to the 'soul', it visualizes the 'sole' of the foot.

THE VALUE OF ANALOGY

One of the uses of analogies in everyday life is to make things clear. We say 'it is like', 'it is as if'. Such an analogy is of course no argument at all, although it is constantly used as such, but it helps us to understand a difficult or abstruse problem; and the curious fact is that when we understand something clearly we are apt to take it to be true – 'I can understand *that*!' we say. When we accept an argument by analogy we are following primitive thinking in accepting that there is a real connexion as well as an analagous one, that 'things that are similar are the same thing'.

The idea that two things which happen together in time are

causally related is of course logically false. Nevertheless, this capacity to bring things into juxtaposition is the cause of many scientific discoveries which reason has failed to achieve, for that is the basis of analogy, and analogy leads to discovery. We have referred to such scientific discoveries as the work of the subconscious mind (p. 113). We can now say that it is largely by the recognition of similarities that the subconscious mind is able to make these discoveries.

William James, as we have noted, said that genius is the capacity to observe analogies. As we have seen, it is in relaxed states of mind that the mind works by association of ideas and the recognition of similars; so Newton's discovery came when he was relaxing under an apple-tree and realized the analogy between the falling apple and the world's suspension in space. It was an analogy of snakes swallowing their tails which led to the discovery of the benzene ring (p. 113). There is no logical connexion between these two, but the similarity led to a great discovery. Archimedes discovered his principle concerning the displacement of water and shouted 'Eureka!' when reclining in his bath. It is of course by the observation of similarities that we are enabled to formulate what we call 'the laws of nature', for these are based on the observed regularity of events in nature; that in certain sets of circumstances certain other things invariably follow – for instance, that solids expand when heated, and that ponderous bodies always fall to the ground, an observation which is summarized in the Law of Gravity. When we find apparent exceptions to the rule, that enables us to test or 'prove the rule' – a phrase commonly misinterpreted to mean that an exception in some mysterious way goes to prove that a rule is true.

But such argument by similars rather than by logic may, like intuition, lead us to false conclusions; and that is probably the origin of many superstitions, which regard as cause and effect many things which have no logical connexion whatever. Your shadow is like yourself, therefore it is yourself, therefore you must pull down the blinds when death is in the house, or else death might get your shadow, your soul. Analogies may be pressed too far: they help to make things clear; they may enable one to see similarities in events which turn out to have a logical connexion; but unfortunately they are often used to misguide one as to the real issue.

DREAMS ARE ANIMISTIC

Another characteristic of primitive man as of dreaming is that he is *animistic*, and to him all things material are living – trees, sun, and storms as well as man and animals. In primitive life the causes of events are never natural causes; accidents, old age, death are never natural but always brought about by magic powers.[1] There is no such thing as chance, but there is also no such thing as determinism: everything is caused by living agents. In New Guinea, if a tree falls and kills someone, it is a witch who has caused it to do so even though the tree may have been thoroughly rotten. In dreams we find this same animistic tendency, so that even ideas are dressed up in human form, and qualities of character are represented by people possessing those qualities. In other words, in dreams our thoughts, feelings, and attitudes of mind are often personalized, being represented by some person, known or imaginary, who has that quality. Thus our anger may be represented by a man we know who is bad-tempered; our sexuality in the form of a loose woman; our intellectual life by a professor; the healing forces within us as the doctor. More imaginatively our rage may be personalized as a fury, sex as a vampire, and our feelings of revenge projected and personalized as a witch. In fairy-tales even trees take on animistic form and animals take on human form and speech, a trend in human nature which Walt Disney has popularized in his excellent cartoons. The delight that old and young have in these films is largely based on this reversion to an animistic mode of thought. His animals are very human, which is only another way of saying that humans are very like animals. Thus dreams are always playing at charades and acting out ideas instead of putting them in the form of spoken words. Harpo Marx of the Marx brothers represents the non-vocal primitive element in human life with its 'all or nothing' characteristics. He never speaks, but if he wants to light a cigarette he produces a blow-lamp from his tail pocket!

DREAM AS A DRAMA

We come therefore to realize that the dream is a drama in which all the actors represented in the dream are parts of oneself. Our

1. L. Lévy-Bruhl, *Primitive Mentality*, p. 47.

personalities have many attributes; we have our kindly side, our arrogant side, our ingratiating side, our lazy side. These are often in conflict, and therefore create problems in the personality which may be reproduced in dreams. Because of the tendency of the human mind to animism, all these aspects of the personality are personalized, and in the dream may be represented by persons, whether of our acquaintance or an imaginary person of that same character, all having an argument or discussion – as they might in a play. In the dream at the beginning of this chapter (p. 129) the dreamer's own arrogance was represented by his friend who was 'too big for his boots', and whom the patient had to knock out. In the Prometheus dream (p. 85 f.) the woman representing intuition was an imaginary woman who tells him what to do with the Negro representing his primitive self.[1] In the dream on page 133 the policeman represented the patient's moral self or super-ego and the tramp his primitive natural self, and the dream worked out his problem by the interplay of these personalities as in a drama.

So the dream represents a drama in which all the people are *dramatis personae*, representing, as they do on the stage, certain ideas or types of mind, not the people themselves. In a play a man of self-sufficient character stands for self-sufficiency in the abstract, a priggish man for priggishness, a prostitute for the sensual part of ourselves, a parish priest for consideration for others. These are all characteristics we ourselves possess and the dream presents a drama in which all these characteristics in us are represented by people fighting it out, debating the question, playing out the problem in dramatic form and therefore tending towards a solution. The value of a theatre is that we see ourselves as we see others; in a dream we see ourselves as our subconscious sees us. In treating patients, the technique of psycho-drama, in which the patients play out certain roles, is used for much the same purpose.

As the dream turns thoughts, ideas, and emotions into people representing them, so in our dream interpretation we need to

1. Jung, *Integration of the Personality* (p. 103), gives an illustration where the dreamer rides in the railway carriage and by standing squarely before the window deprives his fellow-travellers of the view. Jung interprets this as standing in his own light; obstructing the light from those behind, that is from his unconscious.

reverse the process and find out what those people stand for and so discover what they stand for in ourselves. This tells us what is the nature of the conflict within our soul and what solution the soul offers.

Let us now look at a dream to illustrate the function of the dream as a drama.

This patient was outwardly a kindly, ingratiating, helpful person, who had had his own personality crushed out of him by a childishly jealous father. Fear of his father, especially when backed by his mother, who felt she must be 'loyal' to the father (and thus did an injustice to the son), made him ingratiating but aloof. His problem was that he wanted affection, but he had had to repress that in favour of a detached and pompous attitude which he had taken over from his father. But this attitude left him without friends and, because of the repression of his aggression, with an inability to work. All his conscious efforts to work and to be sociable were of no effect at all. How does he solve that one? He dreams: 'I was in my house and saw two boys looking through the glass, wanting to come in to see my friend Fred. I did not want to be seen by them, but realized that I had been seen and so let them in. They were not certain that they would be welcomed by Fred. I let them in and then there was a girl there whom I embraced – it makes me want to cry now in speaking about it – the idea of her tenderness and of my being welcomed.' The interpretation: his shutting himself in the house and not wanting to be seen represented his detached attitude towards people which resulted from his treatment in early childhood. Fred and the girl he embraced are in fact man and wife, an affectionate couple and friends of his. Fred he had regarded as a kind of father, friendly, as he would like his father to have been, but also detached and aloof like his father. Fred therefore represented the aloof side of his personality. The two boys who wanted to be sociable and friendly represented that other part of his personality which wanted to break down his aloofness and be friendly: the 'two' the intensity of his need for friendship. But he was dubious as to whether Fred, his aloof pompous side, would welcome the expression of his need for affection. However, he does open the door and admits the friendly spirit into his personality. Immediately he does this he is overcome with affection as represented by his embracing the girl. Friendship once released immediately develops into love and affection. The girl represents love and the tender side of his nature, something more than sex and more than mere friendliness which is represented by the boys. This tenderness which he had always longed to express is now released, so much so that in speaking of this affection and love in the dream he is filled with emotion. The dream brings together all these sides of his personality and works out in the form of a drama a readjustment of his personality: it shows him throwing off his detached aloof self, admitting friendship in the first place, and thereby releasing the affection and tenderness in his nature which he had so long repressed. Let it be noted that quite irrespective of any interpretation, the effect of the dream was in fact to release this tenderness: 'It makes me want to cry!' he says.

Freud is inclined to stress the more objective interpretation, Jung the subjective. According to a Freudian interpretation, Fred would stand for his father (who is none too friendly), and the woman he embraces, his mother – an Oedipus situation of a sexual incestuous nature embodying a wish-fulfilment. But the patient was in fact alienated from his mother, since she had let him down by agreeing with his father. If the woman were taken to be the mother, the dream then might represent a reconciliation with his mother, a wish-fulfilment of restoration to his mother. A Jungian interpretation might be that the female form is his *anima*, his femininity and tenderness, the suppressed part of his personality, the opposite of his masculine independence and detachment, and the dream would be interpreted as compensatory. From the practical point of view the upshot is the same – the release of his repressed emotional need for affection.

It would, however, be stressing the subjective aspect of dreams too far to contend that people in the dream always represent something in oneself, although it would be true to say they always have some relationship to oneself. In any particular dream, therefore, we have to decide whether a person stands for some objective person, or whether he stands for some quality or characteristic of that person taken to represent that same characteristic in the dreamer himself. The man dreams of the doctor treating him; this may stand for the doctor himself objectively, or the doctor as a symbol of the patient's father, or the doctor as healer, or the doctor as symbolizing the healing forces in the patient himself. The choice depends on the patient.

Now, it may be asked, if each person represents ourselves, what does 'oneself' in the dream represent? – for often *we* are in the dream, as in the dream just described, the dreams of the arrogant officer (p. 129) and the policeman (p. 133 f.). The answer is that it represents the aspect of our personality with which we *identify* ourselves, our accepted or dominant character, as it were. In the dream just mentioned, the boys, the woman, and Fred all represented parts of himself, but his dominant character, the one with which he identified himself, was the detached attitude and therefore that was the one which is himself in the dream. But his identification changes, for he finally becomes identified with the friendly, loving, and affectionate side. So in the other dream the patient at first identifies himself with the policeman who arrests the naked man, then with the indifferent person who is amused at the policeman washed up by the tide, and finally associates himself with the tramp.

This transition from one identification to another in the dream is obviously of the greatest significance in treatment as showing the progress of the patient and where he now stands. A woman patient with hysterical headache dreams first of a sick woman whom another woman is scolding and telling that she can quite well pull herself together, but the sick woman resents it and gets worse. In this dream the patient is not identified with either party, but both are parts of herself, the one full of self-pity, the othere severe on herself for her self-pity. In the second dream she is herself scolding a woman who was paralysed (her neurotic self) and telling her she is only trying to get out of things. Finally, a nurse comes along, deals with the sick woman sympathetically, encourages her, and gets her to walk. This dream is one in which all three of these women are the patient herself, different phases of her life, one her neurotic self who wants to be ill and is determined unconsciously to stay so; one representing her super-ego which is severe and censorious, but which by her very severity perpetuates her illness: and finally, there is the nurse, representing a more tender attitude of mind towards herself, more sympathetic and less censorious, and this attitude, the dream says, is more effective in curing her. The dream, in other words, answers her question, a question perpetually confronting the neurotic patient: 'What attitude am I to adopt towards my illness? Should I be stern with myself, or should I give in to it? In any case, how can I get rid of it?' The dream gives her the answer. The nurse's attitude is not a mere wish-fulfilment to get sympathy, it signifies the right attitude for her to adopt towards herself, and tells her that she is being too hard on herself. It was because she was too hard on herself that she repressed her self-pity (which her circumstances in childhood fully justified), and it was because she repressed her self-pity that it emerged in the form of her paralysis which was unconsciously designed to get the sympathy she denied herself. The dream encourages her to see that a more sympathetic attitude towards herself would be more effective in curing her.

Thus by regarding the dream as a drama, and the persons in the dream as the *dramatis personae* representing characteristics in ourselves, we may discover the nature of our problems, and by watching the actions of the people in this drama we may discover what our subconscious mind suggests as to a solution of our problems.

CHAPTER 7

FREUD'S DREAM, 'BIOLOGICALLY' INTERPRETED

IN order to illustrate the difference between the 'biological' interpretation of dreams and Freud's 'wish-fulfilment' theory, we may take a dream of his own. Freud gives this dream to illustrate his theory, but it appears to support the biological theory more than that of wish-fulfilment.

The dream is given in *The Interpretation of Dreams*, Chapter 11, p. 89, to which the reader must be referred for further details. We shall, however, give the basic facts of the dream material and accept Freud's own free associations, but we shall suggest a very different conclusion regarding the interpretation.

The facts were that at the time of the dream Freud was treating a personal friend, an attractive young widow called Irma. At a certain point he 'expected her to accept a solution [obviously sexual, since the sexual interpretation was the only one he gave at that time] which did not seem acceptable to her' (p. 88), and in this disagreement treatment was cut short with only partial cure. A colleague, Otto, who knew her, remarked in a tone of reproach that 'she is better but not altogether well'. This worried Freud, who thereupon set about writing down the whole history of the case, 'as though for my justification', to Dr M., a mutual friend, who was at that time a leading figure in their circle.

Freud's dream of 23–4 July 1895 was as follows:

'A great hall – many guests whom we are receiving – among them Irma, whom I immediately take aside, as though to answer her letter, to reproach her for not yet accepting the "solution". I say to her: "If you still have pains, it is really only your own fault." She answers: "If you only knew what pains I now have in the neck, stomach, and abdomen; I am drawn together." I am frightened and look at her. She looks pale and bloated; I think that after all I must be overlooking some organic affection. I take her to the window and look into her throat. She shows some resistance to this, like a woman who has a false set of teeth. I think anyway she does not need them. The mouth then really opens without difficulty and I find a large white spot to the right and at another place I see extended greyish-white scabs attached to curious curling formations which have obviously been formed like the turbinated bone – I quickly call Dr M., who repeats the examination and confirms it. ... Dr M.'s looks are altogether unusual; he is very pale, limps, and has no beard on his chin. ... My friend Otto is now also standing next to her, and my friend Leopold percusses her small body and says: "She has some dullness on the left below", and also calls attention to an infiltrated portion of the skin on the left shoulder (something which I feel as he does, in spite of the dress). ... M. says: "No doubt it is an infection, but it does not matter; dysentery will develop too, and the poison will be excreted. ... We also have immediate knowledge of the origin of the infection, my friend Otto has recently given her an injection with a propyl preparation when she felt ill, propyls. ... Propionic acid. ... Trimethylamine (the formula of which I see printed before me in heavy type). ... Such injections are not made so rashly. ... Probably also the syringe was not clean." '

Freud's interpretation of this dream is that it is a wish-fulfilment. He had in fact partially failed with Irma, as Otto reminded him, and so in the dream he first throws the blame on her, reproaches her for not accepting the solution, and says, 'It is really your own fault', to justify his wish. Then in the dream he further excuses himself for his failure by saying that her condition was an organic one and therefore presumably outside his sphere as a psychiatrist (p. 101). So far, it is true, the dream is wish-fulfilment; he is trying to justify himself by casting the blame on Irma herself. But that is only the beginning of the dream, and in fact there was nothing new in that, still less any unconscious motive, for he was already trying to justify himself the night before when he set out to write an account of the whole case for Dr M., which shows how worried he already was about his failure.

Moreover, if the intention of the dream was to reassure him, it goes about it in an extraordinary way. For though it starts off with an attempt to evade responsibility by putting the blame on Irma, this is immediately met by Irma's (and the dream's) insistence on the seriousness of the condition. 'If you knew', she is made to say, 'what pains I have!' And in fact she looks terribly ill and pale. That hardly looks like wish-fulfilment! It looks more as if the dream is trying to shake the dreamer out of his attempt to justify himself, an attempt not to support but to debunk his wishful thinking.

In that the dream appears to succeed, for he says he is frightened, and it makes him fear that 'after all I must be overlooking some organic affection.' To an unbiased observer that would appear to be casting added blame on *him* for an omission after all. But not so to Freud, for in his attempt to uphold his theory of wish-fulfilment he interprets it as an attempt to find an excuse for his failure, since an organic illness is outside the sphere of a psychiatrist. But it is not in fact outside the sphere of a psychiatrist, part of whose job it is to exclude the possibility of an organic affection. The dream holds him to that, and in fact Freud in the dream accepts that responsibility, for he immediately begins to make a closer physical examination of Irma's throat by the window, and indeed finds there an infection which he had overlooked – another reminder of his failure. Once again the dream refuses to allow him the comfort of his wish-fulfilment.

This again alarms him to such an extent that he calls in Dr M., who with his colleagues Otto and Leopold, as well as Irma, all emphasize the seriousness of the condition. Far from reassuring him and satisfying his wishful attempt to evade his responsibility, the whole trend of the dream, so far, is in the opposite direction, namely to bring home to him the seriousness of Irma's illness and his responsibility for his failure – a direct counterblast to his acknowledged attempt to justify himself; and it does this so effectively that he is alarmed and makes further efforts in his dream to rectify his faults. But as though these direct blows were not enough, the dream goes on to remind him of instance after instance of his past mistakes and failures. The infection of Irma's throat reminded him, in his associations, of another patient with diphtheria (white spots) for whom he has recommended cocaine with unfortunate results (p. 93), also of a young man with hysteria whom Freud had sent on a sea voyage and who caught dysentery, for which Freud blamed himself (p. 98). Far from being reassuring and suggesting wish-fulfilment, these instances, by reminding him of his past failures, are trying to lead him to admit his present one and not justify himself, which, as he says, he was trying to do.

Then there is the reference to Leopold, who (unlike Freud in the dream, who had overlooked some infection) made a very thorough examination of the patient (p. 95). Freud interprets this as criticizing Otto by contrast with Leopold. But why Otto? It was Freud's, not Otto's, carelessness in the dream in missing the diagnosis which was the point at issue, and it is obviously Freud's not Otto's, carelessness which is contrasted with Leopold's carefulness. It is true there is wishful thinking here, for Freud is still putting the blame on someone else instead of himself, but this time by his interpretation. Indeed all the way through *it is Freud's interpretation of the dream which is the wishful thinking,* not the dream itself.

Let us now proceed to the reference to Dr M. By his remark, Dr M. is obviously making fun of Freud's attempt to belittle the seriousness of the condition and therefore of his responsibility. He sarcastically remarks, 'It is an infection, but it does not matter. Dysentery will develop and the poison will be excreted.' According to Freud's associations, this remark recalls an occasion when Dr M. was called in consultation by a general practitioner, and discovered that the patient had albumen in his urine (an indication of a serious disease), whereupon the over-optimistic G.P. replied, 'That does not matter, Doctor, the albumen will without doubt be excreted!' (p. 97). Freud's interpretation of this is that he (Freud) is making fun of Dr M. in the dream in causing him to make such a stupid remark and for not accepting the diagnosis of hysteria. But obviously the boot is on the other foot. It is Freud who, like the G.P., is trying to make light of a serious condition, and obviously Dr M., a very able physician and 'a leader in their circle', as Freud admits, is tormentingly making fun of Freud as he had made fun of the G.P. He is obviously ironical at Freud's attempt to evade the issue as though to say, 'Irma has all these serious complaints – but it is of no consequence, it will excrete itself. It is her own fault anyway!' Which was more or less the attitude Freud was trying to take up. Freud completely misses the irony and subtle humour at his expense, and overlooks the obvious fact that Dr M. is pulling his leg for adopting the same over-optimistic attitude as the G.P. had done in an equally serious situation. It is not always easy to recognize sarcasm at one's own expense and Freud was human like the rest of us; but the bantering tone of Dr M.'s remark must surely be obvious to everyone except the dreamer. Freud says (p. 90), 'I smile about the consultation with Dr M.' without realizing that it is Freud himself that Dr M. is smiling at, as he smiled at the G.P.'s attempt to make light of his mistake. Again it is Freud's interpretation, not the dream, which is the wishful thinking.

This chronicle of Freud's previous failures and mistakes cannot by any stretch of imagination be regarded as wish-fulfilment, nor as reassuring. One does not usually bolster up one failure by recalling a number of other failures. Obviously the dream is pointing to an exactly opposite conclusion; it reminds him of all these past mistakes and so acts as a corrective to his attempt to justify himself. It refuses to let him get away with his attempt to evade the responsibility for his failure and is shattering in its criticism.

Let us now turn to the final aspect of the dream. Freud had in fact offered Irma an interpretation of her symptom, this interpretation being obviously a sexual one. The dream makes Otto give Irma an 'injection', obviously a sexual symbol not only on general grounds but as confirmed by Freud's associations, for the substance trimethylamin, says Freud, 'leads to sexuality, that factor which I credit with the greatest significance for the origin of nervous affections, which I attempt to cure' (p. 98). The dream says that the injection was poisonous, and the syringe (the origin of the sexual injection) was also unclean. This apparently casts a slur on Otto, and on the face of it is an attempt to humiliate Otto, to pay him out for reminding Freud of the latter's partial failure with Irma. But Freud himself has made us aware of the fact that we often, in waking life as well as in dreams, project unpleasant things in ourselves on to other people, and this is apparently what he is doing with Otto. For after all it was not Otto who had given Irma an 'injection', but that is what in fact Freud himself had symbolically done. This is a wish-fulfilment in so far as he accuses Otto of precisely what he had himself been guilty of doing, giving an injection, i.e. a sexual interpretation of her symptoms which Irma refused to accept and which apparently led to her breaking off treatment. This was so unpleasant a fact to accept that Freud put it on to Otto, which is indeed Freud's revenge on Otto for his unpleasant remark regarding Irma's cure, and his tone of reproach to Freud about the latter's failure. The dream goes on to say that Otto (p. 97) sends Freud some cordial called Ananas, a word which he says resembles the family name of Irma (p. 98) and therefore we may assume represents her. It has an unpleasant smell of fusel oil, which, according to Freud, stands for trimethylamin and thus for sexuality. In other words, Otto is handing it out to Freud that the whole episode of his sexual interpretation of Irma's symptoms has an unpleasant odour. Taken in this sense, we can appreciate the final words of the dream, 'Such injections [sexual interpretations] are not made so rashly. ... Probably the syringe was not clean.'

This dream, then, is an admirable illustration of what we call the biological theory. How then shall we interpret the dream from the point of view of what we have outlined as the biological function of the dream, namely the attempt to solve a problem? The problem of the dream is as usual here presented at the beginning; this was, on his own admission, Freud's attempt to justify himself and evade responsibility. This was a problem so worrying that it made him settle down to write out the case to justify himself with Dr M., and it was with this worry on his mind that he went to sleep. This then was the unsolved problem which the dream made him face. It resolved it in no uncertain terms: it refused to allow him to get away with so facile an excuse for his failure as to blame Irma, or to say that it was organic, or to make light of it – all three of which he attempted in turn. It impressed upon him the seriousness of the situation by reminding him of many past failures. Far from being a dream of wish-fulfilment, it was designed as a corrective to his wishful thinking. It is an interesting fact that Freud suspects this, although he does not give the dream that interpretation. With true candour he says, 'I begin to suspect that with this I am only trying to shift the blame from myself' (p. 96). Again (p. 102), 'It is as though Otto had said to me, "You do not take your physician's duties seriously enough; you are not conscientious, do not keep your promises" ' (p. 94). In these casual observations Freud is blaming himself, not Otto, for carelessness. 'It is as though I were seeking every opportunity to reproach myself with lack of medical conscientiousness' – indeed in our view that was precisely what the dream was trying to do. But Freud only 'suspects' this, it is only 'as though'; finally he makes the somewhat naïve remark: 'Curiously enough, there are in this thought material some painful memories.' Indeed there are, and many of them! That again is precisely, in our interpretation, what the dream intended in order to bring home its lessons. The dream does not appear to have been altogether unsuccessful in this endeavour. For, lest it be thought that we are somewhat harsh in our interpretation of the dream, we may point out the most interesting fact that Freud himself later came to accept the very points which in our opinion it was the purpose of the dream to convey, and one cannot help wondering, in view of Freud's remarks just quoted, whether the dream may not have had some share in bringing this about. For he had given an arbitrary interpretation of Irma's symptoms, and admits (p. 88) that he expected Irma to accept a solution which did not seem acceptable to her; but later he adds (p. 91), 'At that time I had the opinion, *recognized later to be incorrect*, that my task was limited to informing patients of the hidden meaning of their symptoms' – a mistake which many analysts still make! In other words, he came to accept what we conceive to be the teaching of the dream without realizing that it was that which the dream was trying to impress upon him – that it is a mistake to give arbitrary explanations, and that was the real reason for his failure with Irma. The other point is that later he came to realize that the sexual interpretation of neurotic symptoms was not the only one, and that therefore such interpretations 'should not be made so rashly'.

To conclude: Freud's dream, like all dreams, deals with a problem – in his case the problem of trying to justify himself and evade responsibility for the failure in the treatment of Irma. The dream first presents that problem; then it insists that he must face the unpleasant fact of his failure, and presses this home by recalling past failures; it makes fun of his trying to make light of the situation, and finally it insists that it is a mistake to make arbitrary interpretations as he had done in this case, especially sexual interpretations which may, as in this case, lead to misunderstandings. The dream therefore not only stated the problem, but gave the causes of the problem, and worked towards a solution of that problem. Nor apparently was Freud disobedient to the heavenly vision, to judge from his later views.

CHAPTER 8

DREAM SAMPLES

WE shall now give some samples of dreams common to human nature, to show how different interpretations may be put on the same dream, but how in some cases the different interpretations are not so incompatible as they at first sight appear.

FACTORS IN INTERPRETATION

The trouble about dream interpretation, as the reader will long since have appreciated, is that dreams are mostly symbolic, and symbols may be interpreted according to the preconceived idea of the interpreter and made to mean almost what he likes. This accounts for the divergence of the various schools. Hypothesis, so useful in other branches of science, is not much use here, for there is little opportunity of proving the conclusions we draw, as we can put to the test – for instance, some astronomical theory. Indeed the very difficulty is that if we start off with a hypothesis, say, that all dreams are sexual, or that they all aim at a demonstration of power, we can, by juggling with the symbols, adapt the interpretation to the hypothesis we start with and produce a very plausible case. For one the sword is a phallic symbol, for another an expression of power, whilst for still another it stands for justice.

The interpretation of dreams therefore depends at present very largely on the *interpreter* and his philosophy of life. If he is a physicist or physiologist, he will naturally turn to a physical explanation and find much to support his hypothesis that dreams are due to changes in the brain. If he is trained in a strict school of science and thinks only in terms of causes, he will study dreams from that angle alone and may find the interpretation of the dream in the past experiences of the individual. If, on the other hand, he has a broader philosophy of life, and sees life also as a pursuit of ends and aims, he will find purpose in dreams and, if he is a clinician as well, may look to dreams to see in which direction the personality is moving. All these aspects have to be taken into consideration. Indeed, as we have stated, it is our opinion that in the 'complete' dream there is first the statement of the problem, followed frequently by a reference to the causes of the trouble by the revival of some experiences of the past, and then the dream goes on to work out the problem automatically to a solution. It is not so much that you may regard the dream either from the causative or from the teleological point of view, as Jung puts it, but the complete dream includes both aspects.

But dream interpretation depends also on the *nature of the individual dreamer*. An extravert[1] individual (as Jung says) may be well satisfied with Freudian sexual symbols where an introvert would regard these as a superficial explanation, and would find even in the sexual symbols some deeper meaning. We have patients who from the beginning and without any knowledge of psychopathology have typical 'archetypal' dreams and we have therefore in some cases recommended such patients to go for a Jungian analysis, whereas we have had others who have dreams which leave no doubt as to their incestuous nature and are a gift to the Freudian theory.

One dreamer dreams of a womb as a sexual symbol, whereas to another the womb is a symbol of the unconscious, the source from which springs all life and power, and for each the symbol may be the right one.

Dream interpretation is also determined by the *specific problem* in hand. Thus there are some people who dream almost entirely of the things that have worried them during the day, residues of unsolved problems of apparently trifling significance, whilst there are others whose dreams from the beginning might have been taken out of a book of mythology and deal with deep-seated problems of a more universal and collective nature. This fact that the dream deals with the problem in hand may be an explanation of why the patient being analysed by a Jungian tends to dream in archetypal terms, and so with the Freudian – a fact, incidentally, which naturally makes each school more convinced of its theory. This does not necessarily mean that such dreams are merely suggested by the analyst, but may mean that as the analyst deals with certain problems – the Freudian with sex problems, the Jungian with archetypal problems – it is these

1. An extravert is one whose interest and attention are directed to the outside world, and an introvert is one whose interests are towards the inner subjective world.

problems that tend to emerge in the patient's dream, and when they do they naturally present themselves in symbols expressive of their nature, as one person uses painting, another poetry to express his thoughts and feelings.

Thus in the interpretation of dreams the personality of the interpreter, of the patient, and the nature of the problem in hand must be taken into consideration.

These strictures on dream interpretation and the difference in the interpretation of symbols do not, however, invalidate scientific procedure, and, difficult as is the problem, should not prevent us from trying to discover what dreams really mean. But they do indicate that at present the subject of dream interpretation is obviously in the realm of hypothesis; and while it has been subjected to the first process in scientific investigation, namely observation, it has hardly yet entered into the stage of experimentation. Such experimentation would probably best be done under hypnosis, for in hypnosis we can get direct access to the subconscious or unconscious mind under experimental conditions. Indeed a start has already been made, for under hypnosis a patient has been told to dream of a certain subject, and in the dream has transformed the abstract idea into appropriate symbolic form – an experiment, under control conditions, of symbolization and 'dream work'. It would be interesting, too, to get the interpretations of a patient's dreams whilst under hypnosis.

But these difficulties of interpretation make it all the more necessary, as we have said, to obtain the free association of the patient as well as to know something of the patient's psychology if we are to avoid an arbitrary interpretation of dreams. But even the free association of the patient is not always reliable, for he may be evading some unpleasant issue. Patients later in analysis often revert to an earlier dream, having discovered some more adequate explanation.

ARBITRARY INTERPRETATION INADMISSIBLE

The moral of all this is not, as we have said, to invalidate any scientific study of dreams, but it does suggest that in our present state of knowledge we must above all things avoid dogmatism, for, as Jung has said, 'An arbitrary interpretation of dreams is absolutely inadmissible.' In actual practice we should give all these possibilities of interpretation an equal chance; and just as in medical diagnosis, say of a pain on the chest, our first thought should be of all the *possible* causes before narrowing it down by a process of elimination to the actual cause of the present case, so in dreams we should not make a spot diagnosis in the interpretation, nor interpret a symbol by rule of thumb and in one specific way, as Freud does with his list of symbols,[1] and as 'dream books' do, but consider all the possible lines of interpretation before coming to a conclusion as to which is the correct one in this particular case.

When, therefore, we are presented with any dream we need to determine in the first place how far we are to regard it as having merely some present-day problem, as well it may, or how far it may relate to some unsolved problem of childhood or of the past, whether practical or moral, of which this experience is only an instance or illustration, as in the canoe dream (p. 75). Further, we may ask ourselves whether the dream is merely recalling a fact of our nature which we are refusing to recognize, or is it presenting us with a suggested solution, perhaps by way of compensation? Thus a dream in which we are aggressive may reveal to us a tendency to be aggressive with its unfortunate results, which we are refusing to acknowledge, but which the dream is trying to impress upon us; or it may represent our *need* to be more aggressive with its beneficial results. The rest of the dream will usually indicate which of these is intended. But the two are not really incompatible, for if we are repressing our aggressiveness, there is the need to give expression to it if we are to live a full life, and such a dream may therefore be both calling our attention to a problem and suggesting a solution at the same time. In its simplest terms, the dream is making us realize that the aggressiveness is there and ought to be released for the purposes of life.

If the problem is a deeper one relating to our whole life and character, we then look to the dream and find the cause of the trouble in some experience of early life – and so discover what caused our suspicion, our shyness, or dependence on others. We can see why the young married man is sexually impotent when he has a dream of sexual attachment to his mother. This is where

1. *The Interpretation of Dreams*, p. 246.

the Freudian technique comes into its own, in that it traces the cause of the symptoms to their original sources. But we must also keep our minds open for the possibility of there being something still deeper than personal experiences and of a more archaic nature in our dreams. We need to decide whether the dream of a wild animal referred to the fact that we had just seen a film of that nature which frightened us, or whether it represented an old fear of father, or whether the wild animal represented some deeper force in the unconscious of which we are terrified, but which at the same time may be integrated into our personality and used for the purposes of life.

Similarly a woman in our dream may refer to a particular woman of the present day with whom we are having trouble, or to one whom we desire, or of whom we are jealous; or the dream may refer to some complex regarding women as a whole. If so, we must decide whether she represents a sexual object or represents the mother in the sense of a mother complex, which may represent a need for protection and security; or over-possessiveness of the mother, or a desire to break away from her. Finally, the woman may stand for the *anima*, the feminine principle in the personality. Again, if a man dreams of Samson and Delilah, it may be that he has just seen the opera; but the probability is that he dreams it because it touches on some personal problem. Is the dream, then, just a warning against seductive women, or are we to interpret the dream subjectively as the conflict in the man's mind between his ambition and sense of power, as represented by Samson, on the one hand, and his sexual desires, as represented by Delilah, on the other? Or does it mean that a man who over-exaggerates his masculinity and represses the feminine element in his life, the Delilah *anima*, needs to yield to her caresses, that is, to accept the feminine in his nature, if he is to be an integrated personality? It may be any or all of these, but an unbiased study of the dream and its circumstances will usually provide us with the right interpretation.

COMMON TYPES OF DREAM

With these considerations in view, it might be as well to consider some common types of dream with their suggested interpretations.

(a) *Lake or Water*

Let us begin with a common type of dream, namely that of a lake, or water, of the dreamer going down into the sea. Such dreams may be taken in Freudian fashion to be an incestuous desire to return to the mother's womb (to the uterine waters), or it may mean a return to the womb of the collective unconscious from which all creative life springs in order to get a renewal of strength. Or it may refer, as with one patient, to the fact that she once attempted suicide by drowning – a purely personal reference. Even so, it might be maintained that her attempted suicide symbolized her longing to return to the mother's womb as symbolic of protection and security. Or it may refer to the uterine waters, and so to birth, as in mythical stories of heroes like Moses, who were born of water or out of the sea, and in the saying of Christ to Nicodemus that we must be born of water (that is physical, uterine birth) as well as spiritual rebirth.[1] But going underground or into a cellar (p. 82) or into the sea often has reference to the analysis in which it is necessary to go down into the unconscious to discover the causes of the neurosis. Any of those interpretations may account for the frequent recurrence of such material in mythology as well as in dreams: we need to descend into hell before we ascend to heaven; Dante needed to go into the inferno of his own unconscious before he could discover and understand the meaning of life; Ulysses was sent by Circe to the Halls of the Dead to get the necessary instructions before he could proceed in life's journey to his long-lost home. There he would find Persephone the Dread, but also the blind Theban prophet 'whose understanding even death has not impaired.

1. The saying of Christ to Nicodemus that a man must be 'born of water and of the Spirit' has caused controversy amongst theologians, particularly in regard to baptism. But on a broad issue the interpretation appears quite clear. Christ was contrasting physical and spiritual birth. Nicodemus had said, 'How can a man be born when he is old? Can he enter the second time into his mother's womb?' Jesus replies in effect that Nicodemus is referring to physical birth, but he himself to spiritual birth. It is true, he says in effect, that we undergo physical birth, but 'That which is born of the flesh is flesh; and that which is born of the Spirit is spirit.' A man needs not only to be born of water, that is the uterine water, physical birth, but of the spirit – spiritual rebirth (John iii: 4, 6).

For deaf though the prophet was, Persephone has left him a mind to reason with.' Thus Homer recognizes the dangers in the unconscious, but also the functions of reason even in the unconscious, blind though it is. These differing interpretations, incompatible as they may appear, are perhaps not quite so contradictory as they seem, for the descent into the unconscious is a necessary prelude to further progress in life, whether we regard it as the need to seek the cause of the trouble or for a renewal of strength.

(b) *Exhibitionist Dreams*

Another illustration of the same kind is the following, which is a common form of dream. A woman whose symptom was blushing, dreams: 'I am walking down Regent Street with no clothes on, but I am surprised to find that I am not embarrassed and nobody else seems to notice.' Is such a dream merely a desire for exhibitionism, a wish-fulfilment as Freud would say? Or is the exhibitionism symbolic of a need in her life to be more open and frank and not so secretive – a more constructive interpretation? These interpretations, we suggest, may not be incompatible – indeed they may be causally related, for her present-day secretiveness may be, and indeed very commonly is, due to an act of perhaps quite innocent exhibitionism in her childhood for which she was scolded and made to feel ashamed, with the result that she thereafter repressed her showing-off propensities and suffered from general self-consciousness, reserve, and secretiveness in every aspect of life. The dream would then express both the original repressed wish and also the present-day need. Nor is this merely to say, as Jung says, that we may regard the dream *either* causally *or* constructively. For the two are in this case causally related in that the wish for exhibitionism, being repressed, gave rise to the shyness and reserve and therefore the present-day need to be more open and frank. The function of this dream is in the first place to call attention to the latent causes of her neurosis (blushing) by referring to her exhibitionism, which in childhood started the trouble; that is the causal aspect of the case. On a 'biological' interpretation the purposive side of the dream was trying to solve her problem of shyness and blushing by getting rid of her early shame of exhibitionism, and it does this not merely by giving expression to her exhibitionism in the dream, but by reassuring her that 'no one minds, no one takes any notice'

– a rectification of her attitude which is in direct contrast to her early childhood experience. By making her exhibitionistic, but telling her it does not matter, the dream quite automatically gets rid of her embarrassment and restores her self-confidence; and the dream does this independently of any interpretation, merely by her undergoing the experience – she 'sees for herself' that nobody minds.

In such a dream all these theories may be coordinated, for the dream reveals the causal aspect of her symptom, her exhibitionism, which Freud emphasizes, and which causes her subsequent reserve; it compensates for this by expressing a frankness and naturalness which the patient sorely needed, and it demonstrates how a dream reveals the cause of a trouble and automatically works towards a solution. In all theories it is the release of the repressed tendencies which is the central theme and purpose of the dream.

(c) *Cave or Cup*

According to Jung (*Integration*, pp. 140–3), another Archetype is that of a cave or cup. In mythology, as in dreams, this cave often has springs emerging from it (p. 143) and a serpent. It is a cave at Lourdes from which springs of water arise which are said to have the power of mystic healing. To a Freudian the sexual symbolism is obvious – the cave being the vagina, the snake the phallus or male sex organ. But in mythology the snake represents several ideas, apart from the sexual. It often represents *healing* as in the temple of Aesculapius, and as reproduced on the badge of the R.A.M.C., and also in the story of Moses who lifted up the serpent in the wilderness so that all who looked upon it should be healed of the disease which had stricken them. The serpent in mythology also represented wisdom, as in the temple of Delphi. The fact that snakes symbolize healing and wisdom seems to dispose of the idea that they are *merely* sexual symbols. But it is possible that these interpretations are not incompatible, for sex has a stimulating, exciting, life-giving effect. Sex may also be the symbol of wisdom, for it is the sign of maturity and reaching years of discretion, as in the case of Adam and Eve who ate of the tree of *knowledge*, which was obviously sexual. (Adam excused himself on the grounds that Eve tempted him with the fruit, which is a common symbol of the breasts, and Eve retorted that she was tempted by the 'snake', that is by the male sex organ.) Therefore it may be natural that a symbol which starts off as sexual as in the Freudian interpretation may come to represent other aspects of life such as healing, maturity, and wisdom. But the dream of the cup or a cave may also have a personal reference. A patient of mine dreamed: 'I had a goblet like a communion cup which was old and made of wood. Then it began to crack, and when it cracked and powdered into dust it turned into gold.' This she connected historically to an occasion when her father compelled her to admit what was not true, which broke her spirit (crack of the goblet – the goblet being a symbol of her whole personality). But as she proceeded in free association, she sees 'my golden cup and my father puts his nose into it to see the contents, which he regarded with some contempt' (he was always poking his nose into things and he often expressed his contempt for her personality). But now in waking dream associations she feels she does not mind in the least what he thinks. I told the dreamer (who was undergoing a 'training analysis') what would probably be the Freudian interpretation and also the suggestion of the cup as being an archetype representing the unconscious. But she said 'No!' and insisted that her associations were that the cup represented her *personality* which 'contains' something, namely herself. This had been broken by her father's treatment of her, but the dream reassures her that she will develop a richer, more valuable (golden) personality and will be indifferent to what her father thinks. The search for the philosopher's stone which can turn baser metals into gold represents an experience common to us all, namely the search for that wisdom which can transform the crude material of the primitive instincts into something more valuable – which Freud has called sublimation.

We now come to other examples of dreams which are common to many people, such as that of 'things coming to life', failing to catch trains, or just catching the trains, of being chased but rooted to the spot, of teeth falling out, of flying, or of the death of a loved person. Whether these dreams are common because they relate to common individual experiences or problems, or whether they are common because they are archetypal or instinctual is yet in doubt. Freud regards them all as sexual – and they may be sexual in any particular case – but all are capable of other explanations.

(d) *The Dream of being Chased and being Rooted to the Spot*

This may be interpreted as the excitement and 'paralysis' associated with a sexual orgasm. But if we trace such a nightmare – for such it usually is – to its source, we frequently find that it originates in a terror experienced in infantile life; for an infant in a state of fear has the impulse to escape but lacks the mobility to do so, since it has no power over its legs. It therefore feels rooted to the spot, as indeed it is. It is a dream of insecurity, not necessarily of sex, although it may later come to be linked on to sex, especially as sexual excitement is also, as we have said, associated with rigidity and immobility. There is then a conflict between the *desire* for such an experience because of its sexual association, and a fear of it because of the infantile experiences.

(e) *The Dream of Things coming to Life*

This is a projection in objective form of an inner subjective experience. There are in all of us latent potentialities which lie dormant until they are awakened to life by some experience; this is the process of maturation. Interest in art, in music, in religion, which, as we say, 'opens up a new world to us', are potentially present in us, and may be awakened by some passing stimulus. The process of maturation in adolescence brings to life a world of romance, arouses new potentialities of love which change our whole life. A youth or girl to whom we are recently indifferent becomes the most beautiful thing in the world – transforms the whole of life so that everything in life looks lovely, not because they have changed, but because new emotions have sprung up within *us*. This coming to life of latent powers from within is of course a common experience of mankind and is embodied in the myth of Pygmalion in which the sculptor falls in love with his own creation which, as a gift from the gods, comes to life.[1] It also appears in dreams such as that already mentioned of the patient who dreamt that he was in Trafalgar Square and all the stone statues came to life – lions, Nelson, and all.

(f) *Dreams of Falling*

Many people experience this when they are just falling asleep,

1. Bernard Shaw made the girl fall in love with her creator, which is commonplace and misses the whole point of the story.

and wake up with a start. It is significant that we speak of 'falling' asleep. As has been mentioned, such dreams probably relate to a fall in blood pressure as we 'drop off' to sleep. Or such dreams may revert to our archaic ancestral fear when we lived in trees and falling was one of our chief dangers. But why then do we not all have dreams of falling? We may assume that whilst we may all have a predisposition to a fear of falling, it is only those who have experienced a bad fall in childhood (or as airmen) who suffer from such phobias. On the other hand, we usually acclimatize ourselves to such fears in childhood unless they are reinforced. Therefore many people have falls in childhood but do not suffer from the phobias or have dreams of falling. In such cases it is frequently found that the fall was perpetuated to symbolize a fall of another kind, a moral lapse (we talk of 'fallen' women), fear of a failure in business, or fear of a fall from the pedestal of ambition or of pride, for the man who is conceited or is over-ambitious justifiably fears a fall. Indeed there is frequently a close connexion between the two in a reverse direction, for if we go far enough back we often find that what made such an individual become over-anxious to be successful was the need to build up a barrier against objective fears in early childhood; but at the back of his mind there is always the threat of a return of those fears should he fail in his attempt, and when the barriers break down, all the fears of infantile life surge up. This results in the typical 'nervous breakdown' of the successful business or professional man.

(g) *Dreams of Flying*

These may be the desire to escape, but more usually the desire to surmount the limits imposed by nature. Other people explain them as ancestral reverberations of our remote ancestry when our ancestors floated aimlessly and without effort in the sea. As we all carry about with us reminders of that life (an instance of which is the Eustachian tubes connecting the mouth to the ears, which were originally the gills), there should be nothing surprising in the possibility that there should remain within us some predisposition to the muscular movements enjoyed in swimming which give rise during sleep to the sensation of floating; and similarly with flying. But it is also natural that the dreams of flying or floating should have sexual significance since sex is associated with an ecstasy of sensuous feeling, 'light as air'. The term 'halcyon days' refers to the lovers destined to float in one another's embrace over a calm, warm, and tranquil sea.

One of the main problems of life is undoubtedly that concerned with *growing up*. There is the urge to grow up; there is also the urge to cling to the protection and security of infantile life. 'The libido lingers but time does not stand still', as Jung puts it. Many neuroses result from this conflict – indeed it may be said of all neuroses that they are infantile reactions to life, designed to excuse us from facing the problems and difficulties of life. This problem is reflected in many dreams, like the following.

(h) *The Dream of Teeth Falling Out*

This is a case in point. Freud regards it as representing loss of semen – a sexual wish. In women, according to Jung (as quoted by Freud),[1] such dreams indicate parturition. Both explanations are possible. But there is a simpler and more direct interpretation, for the time of life when teeth fall out is in early childhood when the milk teeth give place to permanent teeth. It therefore represents a growing up, a passing from infancy to childhood. Many mothers indeed keep their children's first teeth as a symbol of their babyhood which the mother is reluctant to see pass. This dream of teeth falling out in an adult therefore symbolizes the process of growing up, and may indicate that he *is* growing up, or that he *needs* to grow up and 'put away childish things'.

(i) *The Dream of Catching or Missing Trains*

This is concerned with the same problem – that of progress. The train is first and foremost the means by which we travel and is therefore an apt symbol of progress in life – and that is what it commonly symbolizes in dreams. Freud's interpretation is that a train means to depart, that is to die; and the dream of missing the train therefore means 'compose yourself, you are not going to die [to depart].'[2] The trouble about this interpretation is that such dreams of missing trains certainly do not compose us but usually worry us into wakefulness. On the other hand, the dream in which we manage to catch the train usually gives the dreamer a sense of satisfaction, not of dread, as it probably would if it meant departing in death. Such a dreamer usually remarks, 'I

1. *The Interpretation of Dreams*, p. 234. 2. *Ibid.* p. 232.

just managed to catch the train!' which implies successful effort, not fear – or are we to assume that it means that the dreamer welcomes death? No! The simple and more acceptable explanation is that the train journey denotes progress in the journey of life, not departure for death, and the person who has these train dreams is one who is bothered by this problem of his inability to progress in life, of his arrest in development, of his need to grow up. We often say of a person who is always 'missing his opportunities' that he 'missed the boat'. Dreams take a figure of speech and turn it into reality. Such dreams therefore represent one of life's basic problems, that of the conflict between progression and regression, between a forward movement towards life and a backward shrinking from life, and they tell us a lot about the state of mind of a patient and of his progress. It may be said that the dream of missing a train is because the dreamer is *too anxious* to press on. True, but he is usually too anxious about getting on because he feels himself to be held back in his progress. In the course of treatment, or as the patient spontaneously gets better, the dream of missing the train in fear therefore passes into the dream of catching the train with pleasure and a sense of achievement, and even, as in a case previously suggested, into a dream of driving the train.

(j) *Breaking Away from Authority; being Independent and making our own Decisions*

There is another aspect of this process of growing up. In childhood the authority is usually the father, and therefore the rebellion is against his authority even to the extent of killing him (in dreams) or, less drastically, of the father dying. That is why girls also have such dreams of killing the father. Freud relates such dreams to incestuous desires to kill the father and to possess the mother. Against the Freudian view is the fact that in those matriarchal communities in which it is the maternal uncle and not the father who has authority over the children, the resentment of the sons is against this authoritative uncle, and not against the father who shares the mother's bed. A simpler explanation is that the dreamer hates his father and wants him out of the way anyway because he stands in the way of his independence, not of his sex. It may be objected against both interpretations that the dreamer as often as not is devoted to his father. But that indeed is part of the trouble – he wants to grow up but cannot break away from his parental attachment and this creates a conflict. The dream of rebelling against the father, or symbolically of a man assassinating the king, represents the rebellion of a son who has been too much under his father's dominance whether out of affection, dependence, or from strict authority and fear, and the only way to break away is by getting rid of his father, which is only another way of saying getting rid of his father complex. This alters the whole complexion of the dream. For the dream is no longer a morbid expression of an infantile wish but a healthy revolt against an authority which is crushing his life and individuality. The dreamer needs to make his own decisions and take responsibility for his own actions. That is the positive, practical, and purposive function of the dream. But 'the father' might represent the inner authority of our own conscience against which we are rebelling, but the association in the dream will tell us whether our subconscious mind discourages, or whether it encourages a revolt against a too oppressive super-ego. The interpretation we give is of considerable importance to the patient's treatment.

(k) *Difficulty in reaching the Analyst*

We may now mention a common dream which Freud (so far as I know) does not mention, namely that of patients who towards the end of treatment dream of having great difficulty in reaching the analyst. The dream makes them lose their train, the bus breaks down, the taxi meets with an accident, their car is held up by traffic, and they arrive too late for the appointment; and when they arrive they find other people present and they are supposed to do their analysis with others there. Such dreams may be a resistance, but coming towards the end of treatment appear to signify that the dreamer is being subconsciously prepared to break his attachment, his transference, to the analyst, to learn to do without him. The dream does this by putting obstacles in the way of his reaching the analyst.

(l) *Swimming against a Stream*

Let us finally give a common type of dream to show how differently a dream may be interpreted, the simple dream of a man that he is 'swimming against a stream, and finds it requires all his strength to reach the shore'. This may be an atavistic dream, a reverberation of his remote ancestry when his forebears lived in the sea. Or it may refer to a specific occasion when in fact he found himself in such difficulties. Or it may, again, refer to a birth trauma, the difficulty in being born, which may have been derived from a fact of experience. Or the waters might represent, as to a Freudian, the uterine waters, indicating an unconscious sexual desire to return to the mother. Or the stream and water may symbolize the unconscious and the dream refer to the difficulties in which the dreamer finds himself in his struggle with his primitive unconscious forces. Again it may represent a 'masculine urge' (cf. Adler) and the struggle for power and achievement. But the dream may have a simpler meaning, for do we not often speak of 'fighting against the stream' when we refer to a man's struggle against odds, in which case the dream may be factual, having a present-day significance. Or it may be regarded as purposive, for in the end he reaches the opposite shore, thereby encouraging him to struggle on, and, great as are his difficulties, overcome them in the end. An arbitrary choice of any of these explanations is obviously possible; for any one of them is a possible explanation. But it is only by a knowledge of the patient's psychology and of his free associations regarding the dream, aided perhaps by a knowledge of what the symbols usually imply in mythology and in everyday life, that we can discover what the dream signifies in any particular case and for this particular patient. There is, however, this serious difficulty, that if a patient has been taught in analysis to regard dreams from a particular angle, whether as sexual, wish-fulfilment, as an urge to power, or as 'biological', he will inevitably resort to such an interpretation in his free association. The patient's subconscious mind plays up to the analyst. This is also a serious difficulty in the training of psychotherapists, for a pupil, being analysed (as analysed he must be if he is to be an effective analyst) and finding benefit from his analysis, very naturally accepts the theories involved as the only true and adequate ones. That is why the present writer, when Director of Studies at the Tavistock Clinic, London, always insisted that the students in training, by whatever method they were analysed, should attend seminars and discussions led by representatives of the other schools, as well as experts belonging to no school at all.

PART THREE

NIGHTMARES

CHAPTER 9

NIGHTMARES

PEOPLE may ignore their dreams, they cannot ignore their nightmares. For nightmares can be most distressing, casting their shadows throughout the following day. Hamlet shrank from taking his life because he would have 'perchance to dream'. Nightmares are a common cause of sleeplessness, for many people, like the war-shocked soldier or civilian, dare not sleep because of the horrifying dreams that await them. One does not lightly submit oneself to the experience of getting blown up or buried night after night. The night terrors of children are of this type, for not only are they terrifying in themselves, but their effects persist, filling the day with apprehension and foreboding. The child who is frightened by a dog during the day may have a nightmare of the monster, and may continue to be frightened all the next day. The child who dreams of her mother as a terrifying witch will be suspicious and afraid even of her affectionate advances, lest, as in the story of Little Red Riding Hood, the kindly mother turns out to be a destroying wolf.

Frequently these nightmares are the only form of symptom for which the child is brought to the Child Guidance Clinic, and the only symptom of which the adult patient complains, as in the case of the lady who was insulted by a man during an interview in an office in the City on a Saturday afternoon and struck him over the head with a brass ornament. She thought she had killed him, and fled, leaving him on the floor. On the Monday morning she phoned and discovered that he was 'not well', which meant that he was still alive. But she then began to suffer terrifying nightmares of this experience and of the consequences of having murdered him, which made her life a perpetual torment.

DEFINITION OF NIGHTMARES

How to define nightmares as distinct from ordinary dreams is a little difficult: the very origin of the term is obscure. Are we to give that name to any terrifying dream which wakes us up in horror, like the night terrors of the child or the war dreams of the shell-shocked soldier? Or shall we confine it to those dreams in which we are overwhelmed by some monster?[1] Or shall we give the name to those horrifying dreams in which the physiological accompaniments of fear, such as sweating, palpitation, suffocation, and other signs of organic distress, are the most characteristic features?

The distinctive feature of a nightmare in the more restricted sense of the term is that of a monster, whether animal or sub-human, which visits us during sleep and produces a sense of dread. Sometimes it is a witch, sometimes a vampire, which is conceived as a reanimated dead person who returns to suck the blood of living people during their sleep; or a werewolf, a man turned animal, characterized by sadistic cruelty; or it may be a night hag, an incubus, or a mare. The word nightmare originally referred to these monstrous creatures themselves and then came to be used of the dream in which these monsters appeared.[2]

The origin of the word nightmare itself is considered by some to be derived from the Teutonic word *mara*, which is quite different from the word 'mare', a female horse, and dates from the Anglo-Saxon 'mere'. But these words later came to be regarded as identical, and the mare was one of the many creatures which visited one by night.

According to Ernest Jones these animals in the dream represent a human, and that human is the parent, the nightmare referring to infantile sexuality.[3]

He states that 'the malady known as Nightmare is always an expression of intense mental conflict centring about some form of "repressed" sexual desire.' But he adds that this conclusion is probably true of all fear dreams, but distinguishes them by referring all nightmares to the most repressed of all unconscious sexual desires, namely incest. He adds therefore: 'an attack of the Nightmare is an expression of mental conflict over an incestuous

1. See Ernest Jones, *On the Nightmare*, for fuller information.
2. *Ibid.* p. 244.
3. *Ibid.* p. 246.

desire', and that 'it may be evoked by any peripheral stimuli that serve to arouse this body of repressed feeling' – such as lying on the back in sleep. He says later: 'In all cases of anxiety dreams, the latent content represents the fulfilment of a repressed sexual wish, usually an incestuous one, but whereas in the nightmare this always relates to the sexual act, various perverse sexual wishes come to expression in the other forms of anxiety dreams.'[1]

Jones therefore maintains that there are 'three cardinal features of the malady: (*a*) agonizing dread, (*b*) a sense of oppression or weight at the chest which alarmingly interferes with respiration, (*c*) conviction of helpless paralysis, together with other subsidiary symptoms such as palpitation.'

But all of these are the accompaniments of any intense fear: any severe enough dread will affect our respiration, produce sweating and palpitation, and the sense of paralysis as when we say we are paralysed with fear; and this is true whether it be an objective fear of bombs or suffocation, or, curiously enough, a subjective fear of sex or of our aggressive impulses, to which the organism reacts as though there were an external danger. If, therefore, we are to regard the three features mentioned as cardinal features of the nightmare, we must allow for them being caused by other than sexual fears.

In many cases the fear is obviously sexual, such as the fear of the adolescent girl of burglars breaking in on her sleep. Other nightmares symbolically suggest a sexual origin, such as 'riding' the horse, or the witch riding on the broomstick, and in the case of a girl, the incubus 'overlaying her'.

Even so, we cannot agree that the fear is always of sexuality, although commonly it is so. Nor can we agree with Jones when he says that 'conflict of this intensity never arises except over matters of sexuality'. For conflict of an intensity such as to produce physiological sensations of palpitation, oppressive breathing, and the rest can arise from fears of an objective nature, such as arise from suffocation in infancy, from a difficult birth, an anaesthetic badly administered, and not infrequently from being pressed too closely to the mother's breast. We find in fact that these experiences are often the ones which are later reproduced as nightmares, as well as in neurotic symptoms such as

1. Ernest Jones, *On the Nightmare*, pp. 44, 54, and 79.

claustrophobia, and in psychosomatic conditions like abnormal sweating and palpitation. Similarly in adult life, the soldier who has been blown up wakes in terror night after night with nightmares, and suffers from sweating, palpitation, and constricted breathing from the terror; but the fear is not of incest but of the objective experience of danger he has just gone through.

Others maintain that the source of these dreads is to be found in the primeval forces in our collective unconscious which are so powerful that they are personalized as destructive ogres and monsters.

We do not disagree with either of these views as a possible source of nightmares; on the other hand, as we do not find it necessary to resort to sex to explain all these intense fears, so we do not find it necessary to resort to archaic images for many of these monsters, crabs, and vampires, for we find them to be commonly due to the objectification of our physiological sensations, as we shall later demonstrate. They too are created, not inborn; they are individual, not racial experiences, though common to many members of the race, since our physiological make-up from which these fears and sensations arise is much the same.

We therefore suggest that the distinction between an anxiety dream and a nightmare is one of degree rather than of kind, but *we shall use the term nightmare specifically of these anxiety dreams of such intensity that they completely overwhelm the personality; that they give rise to exaggerated bodily sensations of palpitation, sweating, and suffocation, which are indeed the natural accompaniments of intense fear; and, further, that these fears and other emotions like rage are objectified and personalized into living creatures such as vampires or crabs or monsters or witches, and fill us with such horror that we wake up in dread.* But we shall not keep too strictly to this description, since there are all degrees in between the anxiety dream and the nightmare.

But there is another distinction which differentiates the ordinary complete dream from the nightmare and indeed from other anxiety dreams. Both dreams and nightmares are reproductions of unsolved problems. But the difference is that whereas ordinary dreams work towards some sort of solution so that we continue to sleep, in nightmares the conflict is so severe and the terror so great that whilst there is a reproduction of the problem and of its terrifying causes, there is no offer of any solution, with the result that the emotional tension causes such a state of distress that we wake up in terror. They illustrate the point we have already made, that just as we sleep to escape from the problems of the day, so in a nightmare we waken to escape the unsolved problems and horrors of the night.

It is for this reason that nightmares offer us the most direct means of discovering what is the real problem which troubles the individual's life, whereas we look to ordinary dreams to suggest some solution to the problem.

A woman who during a very happy engagement suddenly became depressed for no conceivable reason found the cause to be that on one occasion she had been greatly aroused sexually by her fiancé, and that night had a nightmare in which she saw the ghost of a man to whom she was previously engaged and who was killed in the war looking reprovingly at her. This filled her with a terrible sense of guilt and shame for being disloyal and she woke up in terror. The reason for this emotional reaction was that when her previous fiancé died she vowed she would marry no one else – there was no request for this on *his* part. Now that she felt strong passions towards another, she felt an unconscious sense of guilt and reproach. This made her position hopeless; she was very much in love with her present fiancé but felt the sense of guilt that she should love anyone but her first love. It presented, like all nightmares, an unsolved problem, and she woke to escape it.

Another distinction between the ordinary dream and nightmare is that a dream, though dealing with the same problem, tends to change its form as it gets nearer a solution, whereas the nightmare tends to be repetitive so that we have the same nightmare year after year, all through childhood and even into adult life. 'I still have a nightmare of a black figure at the foot of my bed.' 'I had a constant nightmare as a child, and still have, of eyes staring at me till I wake up in terror.' Such nightmares relate to unresolved childhood problems.

TYPES OF NIGHTMARES

Taking this broad view of nightmares, we shall consider three types:

(*a*) Those derived from *objective experiences* whether in childhood or adult life.

(*b*) Those derived from *fear of our own impulses*, whether sexual or aggressive, or from fear itself, any of which can be objectified as terrifying monsters.

(*c*) Those which are the *objectification of disordered organic feelings and emotions*, which are personalized as crabs, spiders, vampires, and the rest. These three types are of course not mutually exclusive, and the classification is descriptive and clinical rather than logical.

(a) *Nightmares resulting from Objective Experiences*

There are probably nightmares which, as we have explained (p. 89), are merely reproductions of experiences of the past, dangers and situations to which we could not adequately react nor adjust ourselves. Such nightmares may arise from objective experiences of dread even in adult life, such as the war dreams of the soldier or civilian blown up or buried, or who has suffered accidents. But the experience is so terrifying and they can do so little to cope with the situation at the time that these emotions are automatically repressed and dissociated. But the mind, as we have seen, cannot leave such objective problems unsolved, and they persist in coming up when at night our inhibiting forces are in abeyance. They may come up only as little snatches of the original experience itself – nothing but a nightmare of a fall, a bayonet, a horrible face, an empty house; or as in the cases mentioned, the noise, the staring eyes, or the figure at the foot of the bed. The mere glimpse of it is enough; we shut out the rest and wake up in horror. A patient, for instance, a war-shocked soldier, had persistent nightmares of a bayonet, which on analysis proved to originate when he saw the officer to whom he was batman bayoneted through the thigh. These terrifying experiences in infancy are the same as those that produce the phobias, such as *claustrophobia* (fear of close space and suffocation), often due to actual suffocation as by the nurse, or frequently from a prolonged birth with suffocation and the feeling of being shut in; and *agoraphobia*, from separation anxiety when left alone immediately after birth while the mother is attended to. The importance of recognizing the nature of such dreams is that they give a valuable clue to the discovery of the causes of these neuroses, and that by discovering and reviving the original experience the patient can now readjust himself to the situation, which he could not do at the time it occurred. The mere recovery of the experience is often enough to cure the patient not only of the nightmare but of any other neurotic symptoms which may have resulted from the same experience, a method of treatment which Freud called catharsis or purging and abreaction.

This occurred *spontaneously* in the case of the war-shocked soldier suffering from severe headache, who, seeing a crane far above his head, had the thought, 'Suppose it were to fall'. At night he had a terrifying nightmare of the time when he was buried by a shell, and from that alone he recalled the *whole* experience and was automatically and spontaneously cured of his headache. Where dreams fail in their attempt to thrust the whole experience into consciousness, we can complete the process in analysis.

It is possible, we believe, to have such reproductive dreams without there being any subtle symbolic significance in them; they carry their own problems, which are serious enough; they have reference to objective events. They persist because we could not at the time cope with their problems and now have repressed and forgotten them. They disturb our sleep because they are still unsolved. Most of such nightmares come from early childhood, as we should expect: for a child is quite incapable of coping with even the ordinary conditions of life, let alone terrifying experiences, without the protection and security of its parents, and if that is lacking, the child suffers from fears which tend to come up in nightmares. But since the original experience is repressed, dissociated, and forgotten, the child has no way of adjusting itself to the situation. The dream, as we have said, is an attempt of the mind to revive the whole original experience so that it can be dealt with, but if this cannot be done, the dread recurs as a terrifying and repetitive nightmare.

But let us take an illustration of a nightmare relating to an infantile experience of an objective nature which might with some justification be regarded as sexual symbolism but which in fact related to an actual traumatic experience. The dream is particularly interesting also on account of the origin of several of its details.

The patient at the age of one and a half was left on a table after a walk, together with some bluebells, which he ate. The nurse came in, was horrified, gave him an emetic without effect, and finally the doctor passed a stomach-tube. He was violently sick and suffered from terrifying shock. These facts were recovered during the analysis. Thereafter he had a persistent and horrifying nightmare all through childhood of 'a pole with slushy babies at the bottom of the pole'. The dream continues:

'I was walking along a lane, a path, and on each side of the path there were some little quivering objects like daisies. Everything in the dream was in a frightful state of rigidity so that everything quivered like a gas jet; and over the whole thing a thunderous atmosphere and foreboding, and I knew that something dreadful was going to happen with this dreadful foreboding. Then the impossible catastrophe happened, and the whole thing bulged up and then all went through an inconceivably small space and came out in the dream on the other side – that was the frightful moment when there was that bulging up.'

This dream might be passed over as a lot of nonsense, as indeed it appears to be. An arbitrary symbolic interpretation might take it that the pole was a phallic symbol, the bulging up an orgasm, and the small space the vaginal orifice. In fact and according to the patient's associations the nightmare related to the horrible experience of the bluebells. The free association of the patient in reviving this incident clearly showed that the 'pole' was the stomach-tube projecting from his mouth and throat, the 'slushy babies' was himself as a baby sick and vomiting. (The multiplicity in a dream means 'very much so' – the qualitative is expressed by the quantitative.) The little 'daisies' turned out to be the pearl buttons on the double-breasted waistcoat of the doctor bending over him as he pushed down the stomach-tube, the 'state of rigidity' was of his body, caused by the tube inside him, the 'quivering' was his own trembling with nervous tension, and there was the 'foreboding' that something dreadful was going to happen. Finally, the 'impossible' catastrophe occurred when sickness, stomach-tube, and all came 'bulging up' through the narrow space of his gullet, which was in his nightmare as it would be in fact, the moment of greatest crisis and terror.

Now such an experience would create a very vivid impression on a baby's mind, every detail of which would leave a deep impression upon his visual memory, although of course he would not know what the details meant at the time. It created also a problem, a horror or fear of life, which the infant would find intolerable; but the whole horrible experience would be repressed and forgotten, with the result that it could return only in dreams and nightmares. It is only by the recovery of such an experience by reductive analysis that we can understand the meaning of the details of the nightmare and resolve the basic problem of his neurosis. An arbitrary interpretation of such a dream would get us nowhere: one could give all kinds of poetic interpretations of the symbolism of daisies, of quivering objects, of the bulging, and get no nearer the real cause of his trouble. To interpret as symbolic what is simple fact is to misrepresent the issue and misdirect the patient. But it also illustrates the fact that such an experience may leave a child with an abnormal fear of life, and this fear being perpetuated by the nightmare may give rise, as in this case, to an anxiety state, which took the form of a constant apprehension, even in adult life, that some untold disaster might overwhelm him unexpectedly at any unguarded moment.

Birth experiences are not infrequently reproduced as nightmares: not that the individual can remember being born, but everything one experiences leaves its register on the mind, and in analysis he can reproduce all the physical feelings and emotions he experienced in what can be a most terrifying experience.[1] Nightmares of suffocation sometimes come from experiences of a prolonged birth, which may also, as we have said, produce claustrophobia. Whether the frequent dreams of 'passing through a narrow tunnel or passage' may have reference to something passing down the intestinal tract or to something passing up the vaginal passage, that is to say, sexual intercourse, as in the dream of the palaces (p. 19), we cannot dogmatically state without analysing out each case. But we have most commonly found it to go back to that perilous journey through the vaginal passage at birth, an experience which leaves an indelible impression on the child. Such an experience is repressed and forgotten but is frequently reproduced in nightmares. The patient in analysis revives and relives with terror the whole process, leaving no doubt as to its source. As an instance: 'I have a nightmare of going

1. See 'The Reliability of Infantile Memories' by the author, *Lancet*, 16 June 1928.

through a tunnel: I have an uncanny feeling I am going head first. I have a feeling in my body as though coming through gas, but my brain is not clear because I can't see. I want to rush out of the room. I am definitely drawn through. Everything round me seems to be moving and drawing me with it. I am not being *pulled*; it was drawing, squeezing me into it. I feel I must get out. I felt it was going on for eternity – that was one of the striking things.' After recounting this nightmare, the patient asked in an incredulous tone, 'Can you dream about your birth?' It was spontaneously recognized by the patient as being a birth dream, and by the release of his experience the patient felt a sense of relief and lost both his claustrophobia and his nightmares.

But these objective experiences of early childhood rarely stand alone as a cause of nightmares, for many children have dreads of illness, suffocation, separation, and bad birth without untoward results; for such purely objective fears normally pass away provided they are not reconditioned and provided the child is given the sense of security. But if they are reinforced by later experiences, or if the child continues to live in an atmosphere of insecurity, these early dreads may persist. As an instance of the first, a patient was for a time trapped in his cabin in a sinking battleship with other men but escaped unharmed. He later suffered from claustrophobia, whilst the others presumably did not. Why? Because he was predisposed to such a fear owing to a suffocation at birth which left no immediate symptoms since he got the feeling of security, but was revived by this experience in the cabin. Neither one nor the other experience was enough to cause the neurosis; the two together reinforced one another so that it precipitated the phobia.

But the original dread may be reinforced by another important factor, namely its association with some moral issue. A child is chased by a bull in a field but comes to no harm. Such a fear might in itself produce nightmares, but if it is not reinforced it will probably pass off. But the nurse takes advantage of this and threatens the child. 'If you are naughty, the bull will get you!' The fear then turns from an objective dread to a moral dread, and every time the child has an impulse to be naughty this threat of consequences returns in nightmares which are thus reinforced by moral issues.

Thus we pass from the nightmares which are the direct reproductions of early objective traumatic experiences to those which are of a subjective and moral nature.

(b) *Nightmares resulting from Fear of our own Impulses*

Such a nightmare was that of a boy who lived on a farm in Africa and at night could hear the lions roaring, a source of actual danger but which did not disturb him as long as he felt the security of his parents. But on one occasion he was accused of touching his sister's genitals and was condemned by his parents. Thereafter he suffered from a persistent nightmare: 'There are some lions in a dark place. They are grinning at me, and in order to prove I don't care I say, "Poof! You aren't real!" – and poke them with a stick. They then chase me and terror comes over me as they chase me over a wide open field.' His associations: 'These animals are, of course, my father and mother, who appear to be kindly, but do me down. They are also myself, since I do myself down in the same way as my father does. In my dream I stir them up with a stick as I had poked my sister.' By so doing he had stirred up his sexual passions, dangerous as wild animals, as well as his parents' wrath. The dream was not a wish-fulfilment – on the contrary, it warned him of the dangers of his defiance regarding sexuality.

Such a nightmare has an objective, but also a moral significance, between which there was a causal relation, for without his parents' protection, which he forfeited by his wrongdoing, he would be at the mercy of these wild beasts he heard at night. Incidentally the symbolism of the lions is an interesting case of 'condensation' (p. 23); they stand for the actual lions, for his parents, and for his own passions which he had stirred up.

Such a child is left with an insoluble problem between self-will and fear of consequences, for he cannot give up his forbidden desires because of his self-will: on the other hand, he cannot come to the parent for protection, for that means not only revealing the guilty cause of his fears, but also renouncing his forbidden desires and the end of his pleasures, which he is unwilling to give up. It is because of the moral nature of such fears that the mere reassurance of the parents that the lions would not get him would not be enough to dissipate his fear, for the child knows that the parents are unaware of the true nature of the fears, which is his sense of guilt; and if they knew, they would withdraw their protection. Thus the child is left with the irresistible force of his self-willed desires against the immovable object of his fears of inevitable disaster. He may later forget the cause of his fear, but is left with an anxiety state the nature and cause of which is now unknown and therefore chronic, or with the sense of guilt for which he has ever to propitiate. The child who is told, 'If you are naughty the bogy man will get you,' may thereafter have a fear and perhaps a persisting nightmare of the bogy man although he may have forgotten why. The fear is perpetuated by the persisting desire.

The majority of 'night terrors' of the child are of this moral nature. That is the reason why they often start about the age of three, for it is then that the child first develops a true conscience and sense of guilt. The same causes, as we have said, lie behind these nightmares and the phobias and obsessions of adult life. We fear suffocation or separation or illness not only because we have these early experiences, but because we have been made to feel guilty of self-willed or forbidden sexual impulses of which we may now be unaware, and these objective fears are imagined as the dreaded consequences and punishment of the forbidden desires. The claustrophobic suffers from the dread of being shut in, not only because he was nearly suffocated by his nurse, but because he had been self-willed and this was the punishment. He therefore represses his assertiveness, but the result is that whenever this assertiveness is aroused even justifiably, it brings up in its train the accompanying fear, of which alone he may be aware. The dread of impending disaster which appears in so many nightmares is usually the result of some unrecognized moral conflict. Parents should be made aware of the fact that the imposition of excessive moral standards or of severe threats for comparatively innocent things is a common cause of nightmares, neurosis, and life-long suffering.

Already, then, we have passed from the nightmares which are the mere reproduction of terrifying objective experiences to those in which the fear is of our own impulses, particularly those of rage, sex, and fear itself, any of which can give rise to nightmares because of their real or threatened consequences, such 'consequences' often being borrowed from some actual objective fear.

Further, any of these dangerous and therefore repressed

impulses may become projected, objectified, and personalized into dangerous monsters, often, unlike the lions of the last illustration, of a purely imaginary kind, and symbolizing in some way the emotion itself – a raging tiger, for instance, representing our own rage. Let us give instances of these.

The Personalization and Objectification of Rage. A girl of six, jealous of her younger sister, got into a furious rage with her, went for her, and then was mercilessly thrashed by her father. She determined not to give in, and did not do so till the thrashing made her faint with the pain. She was sent to bed broken in body, her rage temporarily repressed. In her nightmare, 'I saw a great God-Devil coming towards me to destroy me. I screamed out in terror.' A natural interpretation would be that this God-Devil was her father threatening her. That may have been an aspect of the case. But *her* interpretation in free association was this: 'Those things in the nightmare that came to me – God and the Devil overpowering me, were my own emotions of anger and power strangling me and everything. It was almost as if these feelings grew so tremendous inside me that they burst out and that bursting made me feel tiny and overpowered me. I had given myself up so much to my emotions that they became God and the Devil and overpowered me with them.' So her feelings of rage were first objectified, then projected, and finally personalized as hostile destructive powers come to destroy her. The result of the nightmare was that she was reduced to a state of sapless obedience; and she later suffered from a fear of travelling by bus or train, that is, being in the power of a superior compelling force. I asked the patient (to suggest the Freudian theory) whether this monster did not represent her father who had thrashed her, possibly arousing her sex feelings in doing so. She said, 'Definitely not! It was myself: when I had this omnipotent rage I felt *myself* bigger and bigger till I felt I was just going to burst, and then it took the form of this great creature.'

Another nightmare in which rage was objectified and personalized is interesting because of the transposition of the monster of the dream. The monster is first his own rage urging him on to murder: then the monster is the projection of his murderous feelings threatening to destroy him.

This boy was jealous of his baby brother and made a murderous attack on him and tried to strangle him. In the nightmare –

'Something seems to say to me, "Kill him! Murder him! Murder him!" It's so tight in my brain. It keeps on saying "Try it!", "Go and do it", and yet I keep on withdrawing and have the deadly fear that I might. Now I have got him by the throat and start laughing, "See what I have got." The impulse to destruction is as strong as the will not to. The monster has a terribly vicious look, wild! I felt it all through my body just then. Ugh! I felt I was losing control. I glory in the power which I seem to exercise. I delight in a feeling of viciousness, of murderous hate. But each time there is a feeling of fear. It says, "Oh! don't do *that*!" Something else is coming up now. I seem to be *dozing* and I seem to dream that I'd done it – I wake with a terrible start. Ooch! I've done it! I've *done* it! [Patient, in reproducing it in free association, suddenly jumped and shouted out.] Oh God! I've done what I intended to do. Then I seem to see that *I* was being strangled. I seem to see an ugly form of a man telling me to be a good boy, and if I don't he may destroy *me*. He looks very vicious, a monster. He seems to get *me* by the neck. I scream and wake up and have a terrible sense of guilt for all these thoughts I've been having. It seems to me that it is a sort of warning to me to keep off these evil thoughts. [What is that monster? I asked.] It is myself: it is the self I was frightened of urging me to do something I don't want to do. It seems to me this person is more myself, and if I talked of him in the first person it would be more appropriate.' This patient suffered from a compulsion to strangle, which represented the repressed but still active hate. The rage possessing him first turned *him* into a strangling monster, and then turned into a monster strangling *him*.

One could multiply instances of this sort in which the dreamer is in no doubt that the monster is the projection and representation of his own feelings.

Fear is another emotion which can be so overwhelming that it becomes objectified. 'I had a dream of a peculiar animal, a figure which stood out in the dream and it was very terrifying. But it is fear because it stands out like this! [Patient holding her hands as claws in front.] That nightmare started from an attack of pneumonia I had at the age of three and was burnt with a poultice and became delirious. The shock was so great that I felt as though I was going mad, and I put my hands like that [claws] and screamed with terror.' We objectify fear in common language when we say that we are 'seized with panic' or 'held in thrall' by fear. All that the dream does is to dramatize this as it might be dramatized in an allegorical play, in which fear, rage, or sex are represented symbolically by persons.

Sexual Emotions. The objectification and personalization of *sexual emotions* are no less striking in nightmares. Nightmares of being chased, so common to girls in puberty, are of such a nature. They are explained phylogenetically as marriage by capture and psychologically as the sexual desire to be overmastered. That is no doubt possible; but when we reductively analyse out these dreams and nightmares, we often find that these burglars, tramps, and monster creatures represent also their own sex feelings which pursue and overwhelm them. Indeed in many cases the feeling of being overcome by other emotions, such as rage and even fear itself, produces sexual feelings of a masochistic kind, especially in girls at puberty. Fear itself can be sexually thrilling, as those can testify who go on swirling roundabouts at fairs. So can the fear produced by a punishment, and thus what is regarded as the punishment for the forbidden desire may be the means of its arousal – which somewhat complicates the problem.

Terrifying nightmares of the monster type are particularly liable to occur as the result of a sexual orgasm, for an orgasm overwhelms the whole personality. The child has not the same control as the adult over these passions, which therefore appear as overwhelming and sometimes destructive forces coming to overmaster him. The accompanying sensations of palpitation, of hurry, or of paralysis of the legs, so that we feel rooted to the spot, which are so often experienced in such nightmares, are all the physical accompaniments of feelings experienced during a sexual orgasm. A small boy as the result of an orgasm has the fear that 'I shall be done away with, kidnapped, carried off. It was the fear of a big man running off with me at night.' Prior to that it was the fear of a 'big bear coming and hugging me'. In ordinary language we speak of being 'carried away' by an emotion: in the dream this is dramatized as being carried off by a monster.

Before proceeding further with such nightmares we may ask the question: If, as we have maintained, the monsters in our nightmares are essentially our own impulses and organic sensations, why should these feelings be projected, objectified, and personalized into creatures of the objective world? What is the mechanism of such projection? The answer to that question is really quite simple. These feelings and emotions are from the beginning regarded as objective; they start off by appearing to the child as something outside himself. When an infant first touches its toe it does not experience it as part of itself but as another object; when it has a pain in the stomach it feels the pain as something objective, apart from itself – so much so that even in adult life we say we *have a pain in the stomach*, as though *it* is something apart from *us* and located in the stomach; and adult people will sometimes get angry with a toothache and prod a painful tooth to hurt it for hurting them. This is not necessarily masochistic but aggressive. So when a child has sick feelings in the stomach, it is as if the disgust or pain is an entity in itself, an objective thing. It is only a step from regarding these feelings, emotions, and sensations as objective entities to *personalizing* them, for they are active and potent – a spontaneous force acting apart from us, capable of overwhelming us and of which we are therefore terrified. The child, like primitive man, is animistic, with the tendency to personalize not only objects of the outside world but its own feelings and emotions. So also when the child is 'overwhelmed' by a feeling of rage, of fear, or of sexual feelings, which, starting from the sex organs, spread over its body and overmaster him, he naturally regards these feelings and emotions as some monster, and the distribution of these sensations over his body gives shape to the projected creature, making it a vampire, a crab, or a spider, as we shall see.

From this point of view we should not, rightly speaking, call these processes projection or objectification, because these words imply that the experiences were at first regarded as belonging to oneself and were then 'made' objective, whereas, in reality, they are regarded as objective by the child from the start. Only later on does the child regard these emotions as belonging to its own experience.

A further reason for the *projection* of these organic sensations is our refusal to accept these feelings as ours. These experiences are unpleasant and therefore we reject them, or we may regard them as wrong and so refuse to acknowledge them as belonging to us and repress them: we regard them as an outside thing apart from ourselves. 'It is no more I,' said St Paul, 'but sin that dwelleth in me.' Thus the small girl who was called a disgusting, filthy creature by her mother for playing with her faeces found this rebuke so unpleasant that she entirely dissociated herself from her dirtiness and projected it on to her doll, with whom she played 'dirty' games and then scolded *it* for being nasty and filthy – thus satisfying both her forbidden desires by playing the dirty game, and at the same time satisfying the moral disapproval of them. Then in her dreams the filthy part of her was projected and personalized to become a 'filthy tramp who was covered with horrible sores and flies', and who terrified her in a persistent nightmare. The nightmare persisted because the problem between her old and now repressed desires and her moral sense remained unresolved. In another case already mentioned the child turned her rage into a monster because, as she said, 'This gave me an explanation which relieved me of all guilt – that the monster was outside myself that made me do these things. I pushed responsibility for my guilt on to others.'

These then are the reasons, one natural, one purposive, for the personalization and objectification of these emotions.

(c) *Nightmares which are Objectifications of our Organic Sensations*

This brings us to perhaps the most interesting type of nightmares, namely those which take the form of crabs, spiders, and vampires. Jung regards these as archetypes,[1] and certainly we find them widespread in mythology. They constitute a very good test as to the reality of these archetypes. From our analysis of many nightmares of this kind we find these creatures to be merely the projection of our own physical sensations. Take the simplest example of such projection: a small boy of nine going away to boarding-school for the first time was anxious and had no appetite for breakfast. His mother urged him, but he said, 'I can't eat anything, I feel my stomach is full of crabs!' Is this an archetype, or is it not simply a representation of the sensations in his viscera due to his anxiety? The latter is surely the more probable view. A man complains that he has 'ants crawling about his scalp', which are merely sensations due to changes in his capillaries. Strong emotions, especially of fear and sex, have very disturbing effects on the viscera by means of the widespread influence of the sympathetic nervous system. These emotions themselves, as we have seen, may present themselves in the form

1. *Integration of the Personality*, p. 108.

of monsters and raging animals, or they may express themselves *in the visceral disturbances they produce*, such as palpitation, sweating, and changes in the abdominal viscera. It is these which are objectified, personalized, and projected into the form of spiders, crabs, and vampires, and other such creatures.

The objectification of one's own bodily feelings is illustrated in the following simple but repetitive nightmare of a child who suffered from indigestion as the result of feeding from a bottle of cold milk.

'I have the bottle which I enjoy at first and then feel depressed. Then after the bottle I go to sleep and have a nightmare. The nightmare is of squashy suet puddings which all go squelch. It swells and swells. Then I feel a pain somewhere and then I find the pain is in my tummy.'

Similarly the spiders and crabs of our nightmares are the representation of disturbances in the abdominal viscera, which are objectified and projected in this form.

Most of these types of nightmares relate to sexual emotion. This is as we should expect, since the premature arousal of sex in an infant, especially if it results in an orgasm, produces most violent reactions in the abdominal viscera. For Jones's incestuous fears as producing the most intense fears (p. 176) we should substitute the experience of an infantile orgasm as responsible for this. Indeed many of these nightmares are simply the projection of the physiological sensations of an orgasm. The following nightmare of a sexual experience is characteristic of many. This patient, as a child, had felt unloved by his mother and resorted to masturbation as a solace. He had an orgasm and dreamt:

'I had an erection and that gave me a thrilling feeling. I wanted it to go on to more. That made me make more effort and that makes my stomach feel strong. It gives me a feeling of hunger too, but that feeling is pleasant. Then something seems to give, and it all becomes weak. [The aftermath of the orgasm.] It is all flabby and soft in my stomach and my stomach couldn't stand the strain any longer and failed. It was over-excited – so it produced a feeling of disgust and it all turned nasty. There is something at the back of your stomach that seems to turn over and then all that felt clear and bright turns filthy and soft, and the feeling of nausea runs all over you – over your mind too. It's a loathing in your stomach and your stomach seems to be your whole body. I felt I was a loathsome thing all over; it makes you feel disgusted all over, disgusted with yourself.[1] Then I dreamt of a live crab – a nasty, disgusting thing – it was so horrible that I put it out. Then there came an awfully filthy thing like a spider, but with shorter legs and many more of them. This awful filthy thing came towards me and I felt terror of it. I was in an awful extremity. I couldn't move my legs – hardly move an inch and there was the horror, this awful horrible spider near me – if it touched me with one of its legs it would be too awful – I can't explain it because it is too horrible for explanation. Then in my extremity I felt absolutely defenceless – I shouted out to my mother and I wakened shouting.'

We have in this nightmare the complete correspondence between the organic sensations and the figure of the crab and the spider. The sexual emotions are experienced first as pleasure, then as disgust and nausea, and then projected as a crab and spider. It also answers the question why it is that such nightmares take on the form of crabs and spiders. The crab is the objectification of the gripping sensation due to the contraction of the abdominal muscles in sexual excitement, and the spider with its soft squashy body and projecting limbs is an objectification of the soft nauseating feelings produced after the orgasm, starting in the solar plexus and spreading out to other parts of the abdomen and body like the legs of the spider.

Another spider dream suggests the same association with the abdominal viscera. 'I bathe in a pool of water. I wanted to bathe in it, but it was full of horrible things. Then two spiders came out which were dirty and very tired. But in spite of those things I wanted to go in and swim. Then I woke.' Asked for the association, she said: 'Have these creatures the same meaning as snakes? Because I have had a lot of sex feeling lately and have been messing about with myself.' Snakes of course are a phallic symbol and the patient had been wallowing (the pool of water) in sensuality with the aftermath of nausea and tiredness, but with sex desire still unsatisfied.

It is to be observed that though the spider and crab dreams are

1. It is an interesting speculation whether the common custom of regarding sex as 'disgusting' may be due not simply to teaching, but to such experiences of childhood in which sex is accompanied by these visceral changes which produce nausea.

associated with sexual feelings and especially sexual orgasm, they are not themselves direct sex symbols (like the snake), but are the objective representations of the physical sensations in the abdominal viscera resulting from the orgasm.

In some old Chinese drawings crabs are depicted as being located in the region of the solar plexus, which suggests the visceral explanation we have found in reductive analysis.

A not uncommon form of projection is that of the octopus which seizes the patient with its tentacles to devour her. I have in my possession an automatic drawing of a patient naked, terrified, and as though paralysed on a beach, whilst an octopus arising out of the sea slowly and menacingly approaches her to seize her. That such dreams may represent the sexual desire is not to be denied. But we have had other cases in which the child himself is the octopus, the dream referring back to his own desire to seize and devour the mother's breast with a consuming passion which overwhelms him and yerrifies him. (I have retained the typist's error – it is a most expressive word 'yerrifies!') These ravenous feelings being repulsed by the mother are repressed and are then projected and personalized as the devouring octopus which in turn devours him. Note that in common language we speak of a *devouring* passion: does this mean that we devour or that the passion devours us? Nightmares of the vampire are widespread, as are the myths of those blood-sucking monsters. Vampires are completely mythical creatures: they are the ghosts of the dead who return to suck the blood of the living at night in order to nourish the corpse. They must therefore come from our own imagination, either because we have heard of them, because they are archetypal, or because they are the objectification of our visceral sensations. The last are a very common source of these creatures of our nightmares, as we have demonstrated.

We see then that these nightmares have two different sources: either the vampire is the person himself possessed by an overwhelming passion to attack, or the vampire is the personalization of his visceral sensations

As an instance of the first: the infant himself is a voracious creature sucking the life-blood (milk) of his mother, and often in its passion biting the nipple and actually tasting blood. The monster is himself an expression of the child's voracious nature, objectified as the blood-sucking vampire. Sexual feelings are commonly involved in this process, for the breast is the child's first sensuous object and the ravenous feeding at the breast often produces sex feelings and even an erection in the child, which being repulsed are repressed and projected as the devouring monster. This experience, incidentally, is a common origin of sadism in later life.

So too with the nightmares and myths of 'devouring wolves', where the emphasis is upon the same sadistic desires in relation to food. We are the devouring wolves. In the stories of the Werewolf men turn into wolves, this being a reproduction and representation of what happens to a man under the dominance of an overwhelming passion. A girl in a nightmare sees horrible creatures coming towards her 'who threaten to carry me off and *make me like one of themselves.*' In this case the girl is being attacked by these creatures or emotions, but she also feels herself being turned into a monster herself.

We return then to the 'visceral' type of nightmare, of which the following dream is a good illustration, since it provides us with the origin of such nightmares as the vampire, the snake, and the dragon all in one, without resort to 'archetypes'.

'I dreamed of a snake [the male sex organ] which was exceedingly sensitive and delicate to the touch. It was restlessly moving to and fro. It was milky-white in colour. The snake does not seem to be attached to me in any way, nor am I holding it: it is just there, in mid-air! [The objectification and projection of his feelings.] Then fear came out of the end of this snake [the orgasm], and when it came the snake shrivelled up into nothing [end of the orgasm]. Then it turned into a vampire. I felt absolutely ripped open and went off into unconsciousness with this tension hanging over me, at the mercy of some power I didn't understand. Something had to happen and break that spell. I was transfixed. I twisted and turned, but the fear hung over me – that was the dragon, the vampire that hung over me. The vampire didn't do anything, it just hung there, and I was at the mercy of it. There I was lying ripped open with this vampire over me. So I screamed out when the vampire came, and I cried out, "I won't do it again, I won't have anything to do with sex" [a reference to masturbation, which had brought it all on]. So I was squirming and twisting to get away from the vampire.'

This nightmare not only shows clearly the source of the figures of the snake, the dragon, and the vampire, but it indicates also that the fear may originate from the sexual orgasm itself, quite apart from any moral instruction, and therefore indicates how sexuality may be repressed by the consequences of its own premature arousal, a point we shall refer to later.

In this nightmare the snake is, according to the patient's associations, the penis, which is personalized, objectified, projected, and feared. The vampire came out of the penis, and obviously represents the results of the orgasm itself with its overwhelming sensations which hung over him and filled him with dread. The feeling of exhaustion with all the life taken out of him is what a child experiences after a sexual orgasm and is pictorially represented as the vampire with its blood-sucking reputation, for the child feels as if all its blood and strength has been sucked out of it. But the vampire, he said, had also something of the face of his mother whose condemnation of sex had terrified the child. The vampire is therefore a composite figure of the two dreads, but of the two the patient was in no doubt that these monsters were his own organic sensations which themselves were overwhelming and terrifying. Even in the dream the child recognizes these creatures as the products of sex, for he shouted, 'I won't have anything to do with sex!'

In mythology we have the same conception: in the *Arabian Nights*, for instance, when the lamp is rubbed the horrible genie appears – the rubbing being a symbolic representation of masturbation and the genie the objectification of its emotional consequences. The female counterpart of this story is that of Pandora's Box (the vagina) which when opened released all the evils – and hope alone remained.

This explains why sexual feelings, though pleasant, are feared. The fear may arise from previous threats of the mother. But the main fear in all nightmares is the same as any ordinary fear, the fear of obliteration; for in the child an overwhelming emotion is stronger than the child's personality itself, and therefore threatens to obliterate it. It is the same whether it be an overwhelming emotion of sex, of rage, or of fear itself.

When therefore we analyse these 'archetypal' nightmares to their source, we find their explanation in the early individual experiences of the dreamer. In the 'pole' dream (p. 181 f.) the pole was the stomach-tube and the daisies turned out to be the buttons on the waistcoat of the doctor. The monsters and vampires are the projection and personalization of our emotions of fear, rage, or sex, and sometimes of our organic sensations and feelings. The 'evil eye', when reductively analysed, traces itself back to the eyes of someone glaring angrily at the child with the foreboding of evil.

We are led to the conclusion, then, from the investigation of these and many other nightmares, that they are individual, not racial: to explain these creatures of our imagination we do not need to resort to a mysterious racial unconscious; they are personalizations of our own experiences, which then become unconscious.

But if we maintain that these figures and symbols arise from individual experience and not from archaic representations in the unconscious, we have still to explain why these nightmares and myths are so widespread. The reason is, of course, that the physiological processes aroused by the emotions are the same in everybody, and therefore tend to give rise to the same mental representations.

Lack of food in the stomach produces a contracture of the muscles of the stomach and this contracture gives rise to the sensation of hunger. So the experience of a sexual orgasm, especially in a child, produces certain violent changes in the viscera and muscles of the abdomen, giving rise to gripping sensations which are difficult to describe and therefore are described as 'crab-like', or as 'a crab'. The aftermath of such an orgasm – especially in the child as in the instance quoted – is associated with flaccid feelings in the abdomen which are represented as a 'spider' with its sprawling legs and soft underbelly; and the feeling of being washed-out and exhausted after an orgasm with its sense of fatigue and lassitude is as though our blood had been sucked dry by a 'vampire'. The nightmares take on these forms because these creatures are the most appropriate visual representations of the organic sensations and feelings, which are impossible to describe otherwise; and the universality of these images depends on the universality of the physiological changes giving rise to these mental representations. They are just as appropriate symbols as the lion or tiger for feelings of rage. We cannot believe that a child who has never seen or heard of a snake or crab would dream of one. But when a child has had certain emotional experiences producing violent visceral changes and then sees a crab with its claws, or sees a soft-bellied spider, or hears the story of a vampire which sucks the blood of sleeping humans, he recognizes in that creature a visual representation of the experiences he has had, and he thenceforth begins to have dreams of these monsters as though it were they and not the organic sensations which are threatening him. Nothing can represent so well as a crab certain gripping sensations in the abdomen: so when a child sees a crab for the first time he feels '*that* is like what I felt' and is horrified and yet fascinated by it. Since such organic sensations are alike in one person and another, the same symbol is naturally chosen and becomes universalized in myths. When no such creatures exist we have to invent them – hence the vampires, the satyrs, half man half beast (which most of us are); the werewolf, man turning into animal; and other creatures in mythology. This may also explain why children who have never before seen such creatures as snakes or spiders have a feeling of revulsion or fear when they set eyes on them for the first time. It may be, of course, that there are in us innate reactions of repulsion towards animals of such a dangerous type (for even crabs and spiders are dangerous in tropical countries); but equally it may be that these creatures so accurately correspond to what we have felt that we are nauseated by them, as we were by our own feelings and sensations. Because it is difficult to describe physiological sensations, they are much more easily represented in symbols. To an adult person it is 'as though' a crab seized my entrails, 'as if' a venomous snake struck me and penetrated me; to a child even in the waking state, and to an adult person in nightmares it *is* a crab or spider who has visited them. But that a crab or spider is itself an archaic image which in some way exists apart from the individual, and whose potentialities are transmitted in the unconscious, we find it difficult to believe. What are undoubtedly inherited are the physiological organs whose functions are capable of giving rise to such feelings and sensations which are most appropriately described and clothed in terms of certain external objects which in some way, either in shape or function, resemble them. They are archaic if by that we mean ancient and to be found in the most primitive people, and they are archaic in the sense that the physiological conditions on which they are based are not only ancient but inherited, but they have to be aroused afresh in every individual if they are to become nightmares.

This is a subject on which we cannot dogmatize, and our study of these and numerous other nightmares suggests that neither the archetypal images themselves, nor the potentialities of these images as such, are inherited, but that they arise from certain physiological and instinctive processes which *are* innately determined, and give rise to certain mental representations which, being difficult to express, require to find material symbols in order to express themselves clearly and adequately.

Nightmares have a very practical bearing on everyday life and on the production of neuroses. The reader will have noticed in the various illustrations given that when a nightmare reproduces an experience of the day it does so with far more vividness and strength of emotion than was experienced in the incident itself. The exaggeration sometimes, of course, comes about because the dream taps powerful emotions belonging to some earlier experience which has been repressed and has been revived by a more recent experience. More commonly the exaggeration of emotion in the nightmare is simply due to the fact that during the night our inhibitions are in abeyance and the emotions which have been repressed during the day surge up with full force: the pain we have in our half-sleeping state is greatly magnified, an hour's partial wakefulness may appear an eternity of sleeplessness, and the mild fears of the day become the terrors of the night.

The result of this exaggeration has a most profound effect on our daily lives, for it is on account of this exaggeration that nightmares have such a repressive influenee: the angry mother whom the child defied during the day becomes a terrifying ogre or revenging witch at night who reduces the child to abject submission: the orgasm or temper we cannot control appears as a huge monstrous or suffocating creature. The child may not therefore heed the warnings of the day; he is compelled to take note of his exaggerated fear in the nightmare and so to repress the forbidden emotions.

Inasmuch as these nightmares may come from the projection of our organic sensations, we discover the very curious and interesting fact that *there is within the human organism the mechanism for the repression of its own exaggerated impulses and passions* – a fact of great moment, whether from the point of view of psychopathology or that of ethics. It is important for ethics because it suggests that the sense of wrong, anxiety, guilt, and indeed morality itself, can come about not merely from the example, teaching, dictates, and threats of others, but may be inherent, arising from the natural consequences of our own overwhelming emotions. It is important for psychopathology, because nightmares are one of the most common causes of repression of our rage, our sex, and fear itself, and so lead to the production of neurotic disorders,[1] to which we now turn.

1. *Psychology and Mental Health*, p. 275.

PART FOUR
SPECIAL TOPICS

CHAPTER 10
DREAMS AND NEUROSES

In our previous discussion we have had cause to make frequent references to the neuroses. We must now direct ourselves more specifically to this problem and discuss first the relationship of dreams to neuroses, and then how dreams help us in their treatment. Neurotic disorders seem to be much more prevalent than in a former age. But that is probably because they are so much better understood and so much more can be done for them. They include conditions such as anxiety states, like claustrophobia, fear of closed places; fear of illness; hysteria, such as a hysteric pain or paralysis; psychosomatic disorders, like nervous indigestion or nervous headache; sex perversions, like homosexuality and exhibitionism; and obsessional compulsions, like hand-washing, returning to make sure that the door is shut, that the accounts are correct, and a hundred others. The characteristic feature of the psychoneuroses as distinct from organic illness like pneumonia is that they are due to psychological and not to physical causes, they are psychogenetic and not organic.

NATURE OF NEUROTIC ILLNESS

In previous years these disorders were largely taken for granted; nothing could be done about them except to tell the patient to pull himself together when he did not know what to pull together, to tell him not to worry when that was the one thing in the world that he would like not to do, or to give him dope which alleviated the immediate symptoms but rendered him less able to cope with his problems. These conditions were scorned as the Vapours or Megrims, and the patient called 'hysterical' or 'neurotic', although those using the terms had little notion of the meaning of them. We now realize that these conditions are illnesses just as real as pneumonia or pleurisy, and more distressing; that they are more difficult for the patient to cope with as they affect his power of resistance, and that they require more skilled treatment than ordinary organic diseases demand.

When Lady Macbeth had her sleep-walking turn and was worried about the stain of blood on her delicate hand, crying, 'Out, damned spot!', her lady-in-waiting remarked, 'I would not have such a heart in my bosom for the dignity of the whole body', and the doctor, wise man, remarked, 'This disease is beyond my practice.' In those days it was; in these days such a condition is not beyond our practice. We should call Lady Macbeth's complaint a 'contamination complex', of which there are various forms; we know their causes and with any luck we can cure them. Her condition, like all obsessions, was due to a bad conscience. Refusing to purge herself of the real guilt, she substituted the washing of her hands for the cleansing of the stain on her soul – but of course without effect, as it did not solve the real problem. Thus the obsession, like most obsessions, became chronic and repetitive, and because Lady Macbeth refused to face the real issue it came out in her dream and in somnambulism, in which her suppressed conscience compelled her to propitiate for her crime.

The neurosis and the dream have many points of contact; indeed the dream is often regarded as a form of neurosis, and to a certain extent that is true. Both the neurosis and the dream are unconsciously motivated, both are the result of an unsolved problem, and both, although with varied success, are attempts to find a solution to a problem.

It is therefore not surprising that dreams are commonly used in the treatment of neurotic disorders. Since the neuroses are derived from unconscious sources, and since dreams are 'the language of the unconscious', the dream should be able to tell us something of what is taking place in the unconscious mind and therefore reveal something regarding the origin of the neurosis. In fact most psychoanalysts constantly use them for the discovery of the latent causes of these disorders. Dreams form a most useful adjunct to the other methods used for the discovery of lost memories, such as hypnosis, drugs like evipan and pentothal, and free association. In some ways they give us the most authentic account of what is happening in the unconscious, but unfortunately they speak in symbolic language and their interpretation is a matter of contention between the various schools, which therefore have different theories as to the cause of the neuroses as well as different theories of dreams.

MECHANISM OF NEUROTIC DISORDERS

Before we relate dreams to the neuroses, we must look for a moment at the cause and mechanism of neurotic disorders in general.[1] For the sake of clarity and by way of illustration let us take a simple case – that of a man who was paralysed in both legs and had been bedridden for some years. This was a purely functional paralysis – that is to say, there was no organic cause to account for it, yet he was quite unable by any effort of will to move his legs. The symptoms originated in a motor-cycle accident in which he was thrown head-on to the road and suffered concussion. But he had been trained to be tough, and would never give in, nor indeed show any feeling or emotion; so he dragged himself up and with a fractured skull walked to a lorry till he collapsed. Note that in this case, as in shell-shock cases, the patient was a person of good morale, with a keen sense of duty, refusing to give in however severe the circumstances, not the weakling with whom most people associate the neurotic. In other words, any fear, craving for sympathy, or any self-pity were for him out of the question; such feelings were there, but they were in fact all repressed. But though repressed they were too strong to be abolished. So when circumstances were too overwhelming, and especially when he was temporarily knocked out by the 'concussion', all the repressed fear and craving for sympathy surged up. But even so he would not allow it to come up into consciousness as such, therefore it had to emerge in a disguised form as a hysterical symptom. Like the dream, his symptom was unconsciously motivated, for the patient had no idea whatever why he suffered the paralysis in the legs; like the dream, it represented an unsolved problem, the dilemma between his sense of duty on the one hand, and the overwhelming fear and need of

1. For a fuller account of these disorders, their origin, nature, and cure, the reader may be referred to my *Psychology and Mental Health*, or to its abridged version, *Mental Health and the Psychoneuroses* (both published by George Allen and Unwin, London).

protection and security on the other. Consciously he chose the former and repressed the self-pity, but neither of them would give way, with the result that he broke down with a neurosis. Like the dream, the neurotic symptom was an attempt to solve the problem, but in this case by means of a compromise; for his illness gave him a way of escape from an intolerable situation, but without damaging his morale, for no one could expect a paralysed man to carry on! The symptom therefore satisfies both sides of the conflict, but at the expense of an illness. Dreams, as we shall see, can do better than that.

PRECIPITATING AND PREDISPOSING CAUSES

But why should this man develop a neurosis whereas numerous people have accidents of this kind without developing one? For an explanation of this we must go further back, for such conditions always have predisposing causes as well as precipitating causes, which may be either constitutional, the patient being temperamentally over-sensitive or otherwise ill-adapted to face the responsibilities of life, or the predisposing causes may be of a psychological kind, going back to environmental conditions in early childhood.[1] Very commonly both factors play a part, for obviously the constitutionally over-sensitive child will be more affected by the rebuffs of life than the child of the bulldog breed.

Confining ourselves to the psychological causes, we find that the basic problem in all cases of psychoneuroses (and in this we differ fundamentally from Freud) is that of *insecurity*. But since the child's security comes from the protective love of the mother, its insecurity is experienced as *the feeling of deprivation of protective love*.[2] If a child is assured of that protective love it has the sense of security, has confidence to face life, has courage to adventure, can afford to love others, and grow up to be independent, mentally healthy, and capable of meeting the demands of life adequately. Deprived of that protective love, the child feels insecure, is timid in its approach to life, has no confidence to face the problems of life, cannot afford to love others, becomes self-centred, and escapes from life by illness. Feeling left out and unloved, perhaps because of the arrival of another child who gets all the attention, the child reacts variously to life by depression,

1. *Psychology and Mental Health*, Chap. v. 2. *Ibid.* Chap. v.

by fear, by anger, hate, jealousy, self-pity, naughtiness, or takes to sex as a solace for the loss of love. All these reactions are attempts to cope with the problem of insecurity, and all may persist throughout life as abnormal character traits. This sense of insecurity also comes out in dreams, dreams of loneliness, of being abandoned, lost, unloved, or in night-terrors and nightmares, which the child is left to cope with alone but without the resources to cope with them.

These reactions of hate, jealousy, self-pity, and the rest are 'reaction character traits' which may persist as such. But they are themselves distressing, disapproved, and ill-adapted to life. To be perpetually depressed is intolerable; to be jealous brings disapproval; to indulge in self-pity gets you nowhere. They themselves have therefore to be repressed in favour of an attitude more conformable to the demands of those around, which in psychological jargon we call the super-ego. Henceforth the child has to become good, obedient, helpful, unselfish, brave – for only thus can he get the approval he craves. This creates a duality in his personality; on the one hand, there are the native impulses of his ego, his aggressiveness and craving for affection; and superimposed on this is the super-ego or moral standard of being good or brave which a child has to adopt and by means of which he contrives to adapt himself to his social environment and to himself.

This conflict between the individual's native impulses and the demands of society is the most fundamental problem in human life. Our native tendencies are biologically necessary to us to enable us to cope with life, yet conforming to social demands is just as necessary for our safety and security – especially in face of threats to the child. If we have a stable background in the home, we may contrive to harmonize these conflicting forces in our nature, and this we do by *directing these native potentialities* to social ends, by which we find ways to give them expression and yet in conformity with social demands. In that case we grow up without conflict, healthy-minded, happy, and well-adapted to life.

Few of us have completely ideal conditions, and that is why none of us is entirely free from neurotic symptoms; yet most of us manage to keep a certain balance of mind and to cope with more or less success with the problems of life. But in other cases the fundamental conflict is so severe that we fail to harmonize these two great forces in our personality, with the result that the natural impulses of the ego either break out in revolt against the moral demands society makes upon us or those we make on ourselves, in which case we become delinquent; or if the repressing forces of the moral self or super-ego are too strong for that, the native impulses break out by a back door as a neurosis, which is a way of expressing the natural desires without incurring too much the disapproval of our moral sense. If it is the craving for protective love which was repressed, its emergence takes the form of an illness such as paralysis or pain, which we term a conversion hysteria; if it is the sex instincts which are repressed, they are arrested in development and later appear as the sex perversions; if it is the self-assertive, self-willing impulses that are repressed, then the demands of the super-ego may emerge in the form of obsessional compulsions, like the hand-washing mania, or over-conscientiousness as a propitiation, designed to avert the dreaded consequences of our forbidden desire.[1] It should be emphasized that whereas this conflict was originally between the child and its environment, when these demands were perforce accepted by the child, as in the cases quoted, the conflict becomes a conflict *within the personality itself*. This is what Freud calls an 'endo-psychic' conflict; every neurosis is the result of an endopsychic problem. The cause of a neurosis is not the conflict between a man and his environment, it is always a conflict between a man and himself. That is why sometimes quite trifling things will cause a breakdown, for it is the house that is divided against itself which cannot stand. It also accounts for the fact that one man breaks down as a result of untoward circumstances, whereas another well-harmonized person stands up to them successfully. But it is also why, when a neurosis is once formed, changes of environment have very little effect in curing it: for *radical* cure the endopsychic conflict has first to be discovered and dealt with.

These points are well illustrated in our paralysed patient. He was an unwanted child, as he was the third boy in the family. When the mother then had another child, a girl whom she had longed for and doted upon, the result was that the boy felt left out and jealous, tried to drag the baby off the mother's knee and was severely thrashed to make him good. It did, but at the expense of his repressing all his natural impulses, his craving for

1. For the further description of which the reader is referred to *Psychology and Mental Health*.

love and his aggressiveness, both of which he was made to feel were obviously wrong and disapproved. The result was that he must fend for himself, adopt an independent line towards life, and be self-sufficient. He must face life grimly – and alone – yet without that assertiveness which enables us to cope with life. Nor must he want any love. But life under these conditions is too great a strain, the strain being not so much to cope with life itself as to cope with the standards he had set up for himself; his real trouble was not with life but with himself. When therefore the motor-cycle accident occurred, that was the last straw. But it also offered him a way of escape from the demands of his moral super-ego; it gave him an excuse for getting from others the love he had lost and unconsciously longed for from early childhood.

It may be asked what else the child could have done. The answer is that under the circumstances it could do no other, for the child must have security, and if it cannot get it naturally by the love of the mother, it must resort to these subterfuges. The need of the child for protective love is fundamental, a matter of self-preservation, of life and death. Had the child been given love when it needed it, none of this would have occurred. For the child, healthily and well brought up, given love and the sense of security, responds to the mother's love by voluntarily adopting her moral standards, and instead of repressing, learns to *direct* its natural impulses and potentialities to right and useful ends, in which case there is no conflict and therefore no neurosis.

Regarded in still another light, the symptom is an attempt to restore the personality to wholeness by releasing and giving expression to tendencies and emotions, like assertiveness and love, which should never have been repressed. The illness was primarily an expression, although an indirect one, of his need for love and sympathy, and, secondarily, it was his revolt against going on making excessive demands on himself. Our primitive emotions, like the need for love and aggressiveness, are obviously given us for a biological purpose; they should be used for the purposes of life and directed to right and proper ends. If they are repressed, the individual is handicapped in life and functions incompletely. He therefore cannot cope with life and sometimes breaks down. If these primitive impulses are strongly developed before repression, they will not submit to being completely repressed, and since they are not allowed rational expression, they will express themselves in the unnatural form of a neurosis. The neurosis may therefore be looked upon as a way, a very clumsy way, of giving expression to tendencies which should never have been repressed, and therefore may be regarded as a misguided effort to complete this personality.

But let us remember that all these processes, which end in the production of a symptom, are completely unconscious: they are the work of the subconscious mind in its effort to sort things out, but all it can offer is a compromise in the form of a neurosis. When once such a symptom has been formed, the individual is entirely unaware of its causes and its significance, since these have been repressed; he will stoutly deny that he has any self-pity or that he is anxious to escape responsibility, that he has any desire to hurt; nor is he aware of anything for which he need feel guilty and propitiate, although he feels compelled to carry out these acts. The conflicts and the motive for the neurosis are unconscious – only its manifestations as a symptom are conscious. The person with neurosis is therefore in a hopeless state; he cannot solve his problem because he does not know what it is; he cannot give free expression to these repressed emotions and feelings because not only is he unaware of what they are, but he is careful *not* to know what they are. He therefore begins to rationalize: he says he is only sorry for himself because he has a pain, whereas he has a pain because he is sorry for himself; he says he cannot go on because he is paralysed, whereas his paralysis is due to his refusal to go on; he says he worries because he gets sick and gets indigestion, whereas in fact he gets indigestion because he worries; he says he feels a fool because he blushes, whereas he blushes because he feels a fool; he says he has nothing for which he feels guilty, nevertheless he is compelled to carry out propitiatory acts as though he were guilty – it is that which is so maddening! The patient is quite unaware of the fact that he is playing tricks with himself and therefore cannot solve a problem the very existence of which he denies.

But all these things come up in dreams; for the dream and the neurosis deal with precisely the same problem, a problem which has been shelved, but whereas the neurosis deals with it clumsily and by means of subterfuge, the dream does so more clearly and adequately. What the symptom hides the dream reveals. When we give up the struggle, turn our back on the problem in ourselves, and compromise by developing a neurosis, there is something within us which refuses to give up and continues to work behind the scenes to worry out the problem. Where the neurosis leaves off, the dream begins; where the neurosis gives up, the dream takes up; where the neurosis fails to solve a problem, the dream is often successful in arriving at a solution.

The way in which dreams take up the problem which we have abandoned in developing a neurosis is well illustrated in our hysteric patient. His dreams were in three stages as the treatment proceeded. In the beginning he constantly dreamt of the terrifying accident itself, details of which had been completely repressed, forgotten, and dissociated.[1]

Thus our patient first of all had dreams of being thrown from his cycle, of flying through the air, of a crack on the back of the head, of the horror of death by mutilation – and he would wake with horror from such dreams. So far his dreams were simply recovering the dissociated part of his experience, restoring it to memory, so that by reviving the experience he might face up to it, which he could now do better than he could at the time of the accident itself. This recovery of the original cause is indeed the first function of the dream in traumatic cases, as we have seen (pp. 70, 97), as a preliminary to solving the problem and so curing the neurosis.

Sometimes the dream succeeds in bringing back the *whole* memory or the cause of the neurosis and produces a spontaneous cure – which shows that in trying to bring up the lost memory in analytic treatment we are simply following nature's method of cure. A case in point already mentioned (p. 181) was that of a young man with a severe headache after being buried in a trench, who, when out for a walk, saw a huge crane and thought 'Suppose that fell down on me!' The dream of the whole experience of being blown up and buried, and the recollection in the morning of what had happened, completely and *spontaneously* cured him of his symptom, and his headache entirely disappeared. The neurosis

1. This forgetting or amnesia is generally put down to organic changes affecting the brain and called 'concussion', but it is in fact due to psychological dissociation. For the memory of the whole experience can be recovered in such cases of 'concussion' by such methods as hypnosis, the application of drugs such as sodium amytal, and free association, and they are very often recovered spontaneously in dreams.

made him forget all about the experience, as his friends were always urging him to do, and which he had done most effectively, but he suffered as a result. The dream refused to allow him to forget, but, on the contrary, was always endeavouring to bring back the memory, and thrusting the problem forward for solution. Dreams and analytic treatment have therefore precisely the same function, namely to recover the memory of the forgotten experience to enable the patient to discover his problem, which is the opposite of the neurotic process, which is always trying to repress and to forget.

After a series of dreams of the accident itself, and as analytic treatment proceeded, the patient's dreams took a different turn and began to deal with the problems of his childhood. They passed, in other words, from the precipitation to the predisposing causes, and once again these dreams dealt with the facts of the problem itself without any attempt at solution. The dreams were of loneliness, wandering in empty houses, looking for something, dreams of a child, obviously himself, being left by the roadside in a storm whilst his father and mother went on without him. Such dreams were giving expression to his repressed need of affection. By following these dreams we were able to recover all the incidents of his early childhood already mentioned, of his jealousy of the baby, of his feelings of being unwanted and unloved, and deeper down his feelings of deprivation of love to which he reacted first by depression and self-pity and ultimately by the adoption of the attitude of self-sufficiency and toughness which, as we have found, he could not sustain, especially in face of the accident. In reviving these feelings he sobbed and was full of self-pity; in other words, his love feelings were released from repression, and he was greatly relieved of his symptoms.

The third series of dreams were more constructive: they represented the sublimation and redirection of the feelings so released. He first had aggressive and defiant dreams, quite opposed to his ordinary character, but which were the aggressiveness, jealousy, and resentment he had experienced as a child when he felt himself to be unloved, but which he had had to repress. He dreamt of achieving things and of being successful, which represented the automatic transformation and development of his original hate and aggressiveness into self-confidence and will-power.

Finally, he had dreams of his caring for a sick and lonely child and feeling tender towards it. This may be regarded subjectively as indicating a more tolerant attitude towards his previously 'tough' self. In that sense the dream expresses an improvement towards harmonization of his personality and an acceptance of himself. But we may also see in such dreams a *sublimation* of his self-pity and its development into sympathy for others. This in fact actually took place, and illustrates the way in which the release of infantile repressed emotions spontaneously develops into higher forms *as they would have developed if they had never been repressed.*

This indeed is what we do in analytic treatment. People often criticize analytic treatment by saying, 'Analysis is all very well, but what about the synthesis?' If the analysis is successful, such synthesis is not necessary, as it takes place spontaneously when the repressed emotions are released, as in our patient, with the development of the aggressiveness into self-confidence and will-power, and the direction of self-pity into sympathy for others, without any prompting, building up, or re-educating on our part. Analysis is in fact a process of liberation, not of pulling down and building up. The dream is endeavouring, as we are endeavouring in analysis, to bring about this process of liberation, and that is why it is always reverting, in recurrent dreams, to recover the experience in which the emotions were repressed, for the more the experiences are revived and lived through the more their emotions are released.

COMPARISON OF DREAMS AND NEUROSES

A comparison, therefore, of the dream and the neurosis shows that there are several points of correspondence between the two, and several points of difference. They are alike in that both of them usually derive from unconscious causes; they are alike in that both deal with unsolved problems; they are alike in that they are both attempts at a solution of the problems; they are both attempts to restore the personality to wholeness and completeness.

They differ in that while the neurosis tries to solve the problem by a compromise and a disability, the dream refuses to accept such an unsatisfactory solution, and attacks the problem itself directly and explicitly in its attempt to arrive at one more satisfactory. In this task dreams are more successful than the neurosis. Again, they differ in that whereas the neurosis succeeds in making us forget all about it, the dream is perpetually doing the opposite, *reminding* us of the problem, and recalling the experiences which are the causes of the trouble and which, being repressed and dissociated, produce the symptoms. In this respect the dream helps to solve the problem which the neurosis has abandoned by thrusting the problem into consciousness whenever it has the chance. This it does in sleep when the watchfulness of the repressing forces is in abeyance. Thus while our patient still bluffs himself that he is tough and therefore denies any desire for affection, the dream automatically and repeatedly revives the original crash with all the attendant *fear* which he cannot now deny, and in reverting to early childhood emotionally revives his need for affection. In this way the dream takes up the problem left by the neurosis and invites the conscious mind to attend to it and solve it. Similarly the obsessional patient, who must constantly return to make sure that the gas is turned off lest there should be an explosion, has no idea why she must perform this stupid act. But we know that it is symbolic of a fear of her forbidden and repressed sexual passions, of which she is in fact not aware. By the compulsive act she is perpetually trying to avert the consequences of her forbidden desires. The performance gives her some vicarious consolation, but only temporarily, for it does not deal with the real issue, and at the cost of perpetual propitiation. Her dreams go more explicitly to the root of the problem by bringing to light the guilt and often the reason for the fault. The 'infanticide' dream (p. 26) clearly pointed out to the patient not only his fault but the cause of his guilt, which he refused to recognize, and so urged him to face the moral problem instead of shirking the issue. Such people are constantly dreaming of being found out, of fleeing from justice, of being pursued, of being tried in a court of law and being found guilty. The dreams refuse to allow them to take refuge in a neurosis or to salve their consciences by a propitiation.

SOLUTION OF NEUROSIS BY DREAMS

Let us then summarize the therapeutic functions of the dream in the treatment of neurosis.

First, it makes us aware of the causes of conflict by reviving the experiences which originally gave rise to the neurosis. It tries to do this, but commonly succeeds only in bringing up fragments of the original experience, so deeply are these experiences repressed and so unwilling are we to encourage their recall. That is why the dream has to return time after time to the same experience, and why we have recurrent dreams and nightmares, sometimes in the same, and sometimes in different forms. It takes a long time for the dream to convince us or, to put it more accurately, for us to acknowledge the cause of the trouble.

Secondly, dreams have a therapeutic effect in acting as a safety-valve for the release of repressed emotions. Herein lies the value of the patient himself reliving the experience as he does both in dreams and in analysis. Thus a patient, asked about his dream, replied, 'I had a splendid time! – telling my father just what I thought of him.' Such dreams go by opposites because they give expression to thoughts and feelings which in the day we repress: at the same time, by releasing these repressed feelings, they tend towards a cure of that split in the personality.

The *third* therapeutic value of dreams is that they show us up for what we are, for in a neurosis, as we have seen, we repress our natural selves and impose on ourselves a false personality, a *persona* more pleasing and ingratiating, one more acceptable. The dream calls our bluff. The woman prides herself on being the 'perfect mother'; the dream shows her strangling her child from jealousy. It is as well that she recognizes her jealousy. The war-shocked man bluffs himself that he is still a brave man; the dream shows him terrified out of his wits.

But they may also be a constant guide and help to us in our daily life for the recognition and solution of our everyday problems, and we should heed them as earnestly as the primitive savage does, though in a more knowledgeable way. For dreams, like intuition, are often a better guide than reason in the ordinary ways of life. Merely to go over our dreams during the day has therefore a salutary effect even though we may not fully understand them or interpret them.

When we discover the basic cause of the neurosis in the past, we invariably find that the same problem persists to the present day, and this perpetuates the symptom. The dream, therefore, not only reveals to us the original causes of the trouble, but gives us a picture of the psyche as it is at the present day, our present arrogance or guilt, fear or self-pity. It is not simply a problem of the past; it is also an urgent problem of the present day, and we must be brought to face these unpleasant traits of character which we are only too anxious to ignore, before we can be restored to health and happiness.

DREAMS AND ANALYTIC TREATMENT

It is not surprising, in view of the purposive way in which dreams work towards a solution of our problems, that in the treatment of neurosis, as in his interpretation of dreams, Jung should not bother about the causes and complexes, but concentrate on the present-day problem and discover where the dream is leading us. We agree with this in principle, but since we believe that in the case of the neuroses the present-day problem could not be there were it not for problems in the past, we hold that the problem can best be solved, and our aim best achieved, by looking for the cause of the trouble. This indeed is what the dream does, in that it usually starts off by directing our attention to the cause in the incidents in childhood or adult life, and then leads on to suggest a solution of the problem. Since the dream points the way, why not follow its lead and accept all it has to offer, causes as well as solution? It is good to see where the dreams are leading us, but it is also good in the treatment of the neurosis to see what are the causes of the disorders and deal with these, so that the personality may be free to pursue these ends. Otherwise they may remain as pious hopes, visions of what we might be but cannot be, because these complexes of the past hold us back until we discover and release them.

This is what we do in what we call 'direct reductive analysis', a method of treatment described in *Psychology and Mental Health*, Chapter xvi, p. 420. In this method we take the symptoms rather than the dream as the starting-point of our investigation, and by free association on the symptom itself recover the original experiences which caused the symptom, and the problems which underlie it, release the emotions there repressed, readjust our attitude towards them, and direct them to the use and purpose of the personality as a whole.

Our aim in treatment is not merely the abolition of a symptom but the restoration of the personality to full functioning and to mental health. What the symptom is trying to do but fails to do; what dreams are more effectively trying to do by constantly thrusting the problem before us, we can do most successfully in analysis by deliberately discovering and liberating the repressed emotions and so releasing them for the use of life, so that our lives may be complete and healthy.

Mental health, as I have defined it elsewhere,[1] is the fulfilment of the whole personality: it depends upon the full and free expression of all our potentialities, harmonized with one another, by being directed to a common end, aim, or purpose. This makes for strength of character, peace of mind, and happiness of soul. For this purpose all our inborn potentialities like fear, aggression, and sex, with which we are hereditarily endowed, should be used for the biological purposes for which they have been developed, or sublimated to other social uses.

Freedom of expression is necessary for the liberation of our native potentialities, discipline is necessary for their direction towards a common end. Both are required in the development of the child.

By freedom we do not merely mean freedom for instincts and impulses, freedom to do what we like, for that road leads to conflict between the impulses themselves and still more to conflict between the impulses and the personality as a whole, which results in disharmony in the personality; bad tempers, indisciplined sexual impulses, and, if the conflict results in the repression of these impulses, to neurotic disorders.

By freedom we mean *freedom for the personality*: but that can be brought about only by all the native instincts and impulses being disciplined and directed (not repressed) towards the ends and aims of the personality as a whole. By such means all the native impulses are at the command of the personality so that it develops a strong will and character. In such a life there can be no neurosis because there is no repression, and the personality is free to pursue its ends unhampered by complexes, uncontrollable impulses, and paralysing fears, and so finds its fulfilment and happiness in their pursuit.

If dreams can help us, as we believe they can, to give expression to the repressed and unconscious forces in our nature, and

1. *Psychology and Mental Health*, Chap. 1.

by intuitively guiding, harmonizing, and directing these to their right and proper ends, work towards the full functioning of the whole personality, then surely they are worthy of our serious study, and a right appreciation of their meaning and value for our daily lives will recompense us for our endeavours.

CHAPTER 11

DREAMS AND PSYCHICAL RESEARCH[1]

SCOPE OF INQUIRY IN PSYCHICAL RESEARCH

PSYCHICAL research is the name given to the scientific study of the paranormal, that is to say, to those phenomena which cannot be explained by any known laws. The term 'paranormal' is nowadays used instead of 'abnormal' because many abnormal conditions such as the psychoneuroses *are* explainable by the known laws of science. Nor is the term 'supernatural' appropriate, for the phenomena studied in psychical research may be found to conform to quite natural laws although we do not yet know what.

Its main purpose has been to study such subjects as telepathy, spiritualism, poltergeists, to determine first whether the reported phenomena are genuine, and if they are, what is their explanation.

Societies have sprung up in several countries for the study, recording, and discussion of these subjects. In Britain there is the Society for Psychical Research, or S.P.R. as it is popularly called, which has in its seventy years of existence collected together an immense literature in relation to such paranormal material. These cases have been scrutinized and sifted with the greatest care, as might be expected from a society which has amongst its members gifted physicists, mathematicians, psychologists, and some laymen and women who have devoted much of their time to the collection and examination of such material. But even now the evidence is not conclusive and there are severe sceptics as well as strong believers amongst its members.

Let it be said, however, especially in view of the common criticism of psychologists by judges, magistrates, and others, that evidence is daily accepted in a Court of Law – such as that of a woman recognizing a man who had assaulted her in the dark – which would not stand the test of a psychologist's scrutiny for one minute. It is true that unless such very dubious testimony

1. I am indebted for much help in the writing of this chapter to Mrs K. M. Goldney, whose sound common sense and shrewd criticism in the study of psychical phenomena it is a pleasure to acknowledge.

were accepted in a Court of Law, few people would be brought to justice. But the scientific study of any subject, including that of psychic research, has to proceed on a much stricter basis.

We are here concerned only with the subject-matter of dreams, and not of the other phenomena studied in psychical research, particularly as to whether dreams show any indication of telepathy or of foreknowledge and premonition. As this is a mere introduction to the subject, we propose to take only some sample dreams, consider some of the arguments for and against the theories, and leave the reader to judge for himself.

THREE TYPES OF CASE

We shall consider three types of dream.

First, there are those dreams which are apparently telepathic or precognitive, but, strange as they may be, are capable of a simpler explanation.

Secondly, there are those cases in which the dreamer appears to receive veridical information of contemporary events, such as the death of a relative at a distance, not arrived at either by any recognized channel of sense, or by inference from previous knowledge, but which can be explained by telepathy, if we accept telepathy.

Thirdly, there are those strange dreams and experiences, apparently precognitive, which can be explained neither by known laws nor by telepathy, and for which there is at present no adequate explanation.

1. *Naturalistic*

First, a case mentioned in the *Proceedings of the Society for Psychical Research*, viii, 400 (quoted in H. F. Saltmarsh's interesting little volume entitled *Foreknowledge*, London, 1928). 'Mr Brighten, sleeping on board a yacht at anchor, dreamed of a voice warning him of being in danger of being run down by another vessel. He woke and went on deck, but finding everything quiet and in order, although fog had come on, turned in again and went to sleep. The dream was repeated and he again woke and went up on deck. He was rendered so anxious by the dream – and by the fog – that this time he went aloft, just in time to see above the fog another vessel bearing down on him. He shouted to the captain of this other vessel, who put his helm over and thus avoided a collision.' On another occasion 'he was sleeping on board a moored vessel with one companion. He dreamed that the mooring ropes parted and that the vessel was carried out to sea by the current and swamped on the bar. He woke and went on deck in time to see the bow rope part. His companion also woke and followed him on deck, arriving just as the stern rope parted. They managed to hold on with boat-hooks until help arrived and the vessel was re-moored.'

In the explanation of such dreams we do not need to resort to telepathy or precognition, for they can be explained by subconscious perception due to acute sensibility of one or another of our senses. We have already given an instance of this in the dream due to escaping gas (p. 8) and the dreams of physical disease arising from sensations not consciously perceived, but giving a correct diagnosis later confirmed by medical examination (p. 9).

The subconscious appreciation of external sensations during sleep is a well-established fact and can be experimentally confirmed by those cases in which dreams can be produced by the presentation of external stimuli, such as the alarm clock producing the dream of sleigh-bells, as mentioned on page 8. Further, it is generally accepted that a person in sleep can be deaf to all other stimuli but at the same time be *en rapport* with certain special stimuli, as in the case of the mother who is asleep to all other noises but is attuned to the slightest movement or sound from her infant; or the doctor who immediately awakes to the sound of the telephone but is indifferent to the crying of his infant, through which he sleeps soundly. Hypnosis is another instance of the same condition, and has been defined as sleep *plus rapport*, the subject being 'asleep' to everything except the voice of the hypnotist, and therefore obeying it implicitly.

Thus Mr Brighten's mind being attuned to all possible dangers, as any sailor's would be under the circumstances, it is not at all impossible that even in his sleep he sensed the throb of an approaching vessel and this produced the dream of a collision which alarmed and wakened him. This fits in with the fact that dreams relate to a problem and then depict the consequences of an experience (p. 70 ff.). He might have had a 'premonition' or hunch of the same thing in waking life arising from the same circumstances, and sensed danger without knowing why. Fog plays all kinds of tricks with sound, as most of us know who have been on the Newfoundland banks; the same siren gives a blast which sounds miles away and the next blast sounds as if the steamer is right on you. The throb of a vessel can be heard miles away in that quiet sea – and the sea is always quiet where there is fog. As to the breaking of the cable in the second dream, anyone can try for himself and he will find that there is all the difference between the 'feel' of pulling on a rope when the rope gives because it is successfully moving the object, and the 'feel' of the same rope breaking apart. The feel of the cable on which the safety of a ship depends is something to which a yachtsman is ever alive even in his sleep, as the mother is attuned to the slightest movements of her infant although fast asleep to everything else. He can subconsciously feel the tug of the cable even whilst lying asleep in his bunk, and any variation from that sensation might easily be subconsciously appreciated by him in his sleep, in the same way as a person asleep in a train wakes up when the train slows down, or wakes up when the traffic stops at night, although he was not previously aware that he was aware of the noise. Yet he must have been aware of the previous noise in his sleep, or else he would not have noticed the change. In other words, *Mr Brighten's 'premonition' was probably the natural inference from a sensation subconsciously perceived.* This explanation involves us in believing nothing more than that the subconscious mind may sense things which the conscious mind does not, which is a generally accepted fact, and that it can do so in sleep, which is also a proved fact.

Some experiments in 'telepathy' in which I took part, with Professor Gilbert Murray as subject, produced the most remarkable results I have seen, but my conclusion at the time was that this 'telepathy' was due to hyperacuity of hearing. Professor Murray was in the next room (communication with which was only by the passage and two doors, not direct). Some quotation, such as 'Is this a dagger that I see before me?', or a scene, usually imaginary, was decided by his wife Lady Mary Murray, his daughter Mrs Toynbee, or myself. The statement was spoken in a soft voice and written down; Professor Murray was called in, and immediately replied 'Is this a dagger that I see before me?' Accurate results were given in almost all cases. I based my conclusion that it was hyperacuity of hearing on four circumstances:

(*a*) The experiments, I was told, were not successful if he was further away. (*b*) In one case he 'could not carry on because of the awful noise outside' – this noise being in fact the ordinary noise of a milk-cart, not in our own street but in the next. (*c*) The words had to be spoken: and the experiments in which his wife and daughter spoke the words were mostly successful, whereas when I (a strange voice) spoke, the results were unsuccessful. (*d*) In one case the spoken reference was to Thackeray's *Pendennis*, and Professor Murray replied by a reference to his own son 'Dennis' – apparently a case of mishearing. All of this suggests hyperacuity of hearing, although it should be added that Professor Murray was not aware of hearing the references, so that if he heard, they must have been subconsciously perceived.

I am also informed that in other experiments his reply related not to what was agreed upon and spoken, but to something one of the participants had only thought of, but not spoken. If that were so, hypersensitivity of hearing would be ruled out. For further light on Professor Murray's own views the reader is referred to his Presidential Addresses, *Proceedings of the S.P.R.*, xxix, 46–63, and xlix, 154 ff.

II. *Telepathic*

But this explanation of subconscious perception and hyperacuity in these cases cannot account for the following types of dream, the first of which can, however, be explained by telepathy but the second type, being precognitive, can be explained neither by subconscious hyperacuity nor by telepathy, for the events occurred long after the dreams.

As an example of the second type let me take an instance from my own experience – though no doubt the reader can reduplicate such cases – a dream corresponding to a death. A naval rating under my care at Chatham Naval Hospital said, when summoned for treatment, that he did not feel like having any that morning. The reason he gave was that in the early hours of the morning he had dreamt that his brother was killed in France and it worried him. I mentioned to another doctor present that it would be interesting to see whether the dream corresponded to fact. A few days later the patient got a postcard from his parents saying that they had had information from the War Office that this brother had been killed on that same morning in a raid at daybreak.

This dream and other such dreams might be put down to chance, and in support of that it may be contended that thousands were killed in France at that time; and also the thought must often have occurred to the patient that his brother might be killed. True. But take into account that far from being a passing thought or even a common dream with him, this was the only occasion on which he had such a dream; that the death should happen on that very morning, and at very much the same hour; and that this dream should be so vivid and disturbing that the patient could not carry on with treatment. All these facts point to an experience which to say the least is somewhat unusual and suggests something paranormal.

Such a dream, however, can easily be explained as telepathic, if we believe in telepathy, a belief less 'inconceivable' now than formerly, since the rigidly controlled experiments of Mrs Goldney and Dr Soal in England, and of Professor Rhine in America have been published. The paranormal faculty of telepathy is more surely established experimentally than precognition, and involves us in far fewer presuppositions.

Many examples of telepathic dreams of this type – more striking and better authenticated than the instance given – are related in the S.P.R.'s *Phantasms of the Living*, by Gurney, Myers, and Podmore (London, 1886), the argument taking due account of the alternative hypothesis of coincidence; and Mrs Henry Sidgwick's long analytical paper in *Proceedings of the S.P.R.*, xxxiii (1923), is also of immense interest.

If there is such a thing as telepathy, it is reasonable to suppose it would operate on such occasions as that of the dream of the naval rating when the transmitting brain is in the agony of death and therefore in a state of great emotional tension. That is perhaps why these dreams and phantasms at the time of death seem to occur so often.

On the question of how those messages are transmitted (if they are), an attractive theory is that they are conveyed by physical brain-waves, for it is now well established that the brain gives off waves of different strength, shapes, and sizes, and indeed the measurement and recording of these in encephalography is a method of procedure used every day in medical diagnosis. But the distance traversed by these 'telepathic' messages, frequently from the other side of the world, and in the case of the naval rating from France to London, would seem to preclude such a possibility; and there are other arguments against this theory given in Tyrrell's *Personality of Man*, Chapter 7 (Penguin Books, Harmondsworth, 1947). Some other explanation of such transmission less materialistic than the brain-wave theory must be found – but at present the explanation is wrapped in mystery.

Reciprocal dreams may or may not be telepathic. I quote a case from the booklet entitled *Ghosts and Apparitions*, p. 20 (G. Bell & Sons, London, 1938), by Mr W. H. Salter, Honorary Secretary to the Society for Psychical Research.

'Closely related to dreams in which the representation of an incident is divided between two dreamers', says Mr Salter, 'are "reciprocal" dreams, the general idea of which is that two dreamers, A and B, have, unknown to each other, dreams similar in general content, except that the dream of A relates to B, and that of B to A. Reciprocal dreams are notable not because of any correspondence between either dream and external fact – such correspondence may or may not exist – but because of the correspondence between dream and dream. They are rare but have an important bearing on the interpretation of the more common type of dream, the content of which corresponds with external fact. Here is an example. *Case VIII.* Dr Adele A. Gleason writes from Elmira, N.Y., in February 1892:

' "The night of Tuesday, 26th January, 1892, I dreamed between two and three o'clock that I stood in a lonesome place in dark woods. That great fear came upon me, that a presence as of a man well known to me came and shook a tree by me, and that its leaves began to turn to flame. The dream was so vivid that I said to the man of whom I dreamed when I saw him four days later, 'I had a very strange dream Tuesday night.' He said, 'Do not tell it to me; let me describe it, for I *know* I dreamed the same thing.' He then without suggestion from me duplicated the dream, which he knew, from time of waking from it, took place at the same time."

'Dr Gleason fixed the time of her dream by looking at her watch when she woke. The following morning she entered in her notebook: "Night of dream – J.R.J." these being the initials of her friend, Mr John R. Joslyn, Attorney-at-law, whose account, written about the same time, is as follows:

' "On Tuesday, January 26th, 1892, I dreamed that in a lonely wood, where sometimes I hunted game, and was walking along after dark, I found a friend standing some ten feet in the bushes away from the road, apparently paralysed by fear of something invisible to me, and, almost completely stupefied by the sense of danger. I went to the side of my friend and shook the bush, when the falling leaves turned into flame. On meeting this friend, a lady, some days afterward, she mentioned having had a vivid dream on Tuesday morning (a slip for 'Tuesday night'), and I said 'Let me tell you mine first', and without suggestion I related a duplicate of her dream. I was awakened soon after and noted the time from a certain night train on a railroad near by, and so am certain that the dream took place at some hour of same night. This train passed Elmira at 3 A.M."'

It looks as if such dreams are telepathic; no adequate explanation on natural grounds has been put forward.

III. *Premonitory*

We now come to the third type of dream, which cannot be due to hypersensitivity, and cannot be telepathic, since it implies premonition, the dream taking place before the event. I quote from the S.P.R. *Proceedings*, xi, 577, a case which is also mentioned in H. F. Saltmarsh's *Foreknowledge.*

'Lady Q, living with her uncle, who was like a father to her, dreamed that she was sitting in the drawing-room of his house with her sister. It was a brilliant spring day and there were many flowers showing in the garden, over which, however, there was a thin coating of snow. In her dream she knew that her uncle had been found dead by the side of a certain bridle-path about three miles from the house and that he was wearing a dark homespun suit; his horse was standing by him. She also knew that the body was being brought home in a two-horse farm waggon with hay in the bottom. They were waiting for the waggon with the body to arrive at the house. Then, in the dream, she saw the waggon come to the door and two men, well known to the dreamer, carry the body upstairs with considerable difficulty as the uncle was a very tall and heavy man. During this proceeding the body's left hand hung down and struck against the banisters as the men ascended the stairs. This detail gave her unreasonable horror and she woke.

'In the morning, feeling much upset, she told her uncle and

begged him to promise that he would never ride that particular road alone. He promised that he would always make an excuse to have a groom with him when he rode that way in the future.

'Gradually the memory of the dream grew fainter until, two years later, it was repeated in every detail. Lady Q taxed her uncle with having broken his promise, and he admitted that he had occasionally done so. Four years after this, Lady Q, having married and left her uncle's house, was living in London and was expecting her first baby. On the night before she was taken ill she dreamed the dream again with the variation that she seemed to be in her bedroom in London and not in her uncle's drawing-room as previously. She was, however, able to perceive the whole scene as in the former dreams. Then came another fresh point, a gentleman, dressed all in black, whose face she could not see, stood beside her bed and told her that her uncle was dead. She awoke in great distress but, being then so ill, ceased to dwell on the dream. After a few days she was allowed to write a few lines in pencil to her uncle. This note reached him two days before his death.

'During her convalescence, she wondered at not hearing from him, until, one morning, she was told that her stepfather wished to see her. He entered the room dressed in black and stood beside her bed. Lady Q cried out, "The Colonel is dead. I know all about it. I have dreamed it often."

'Subsequent inquiries showed that the dream was fulfilled in every detail, including that of the left hand striking against the banisters. The men who carried the body upstairs were those seen in the dream. The only detail which was not correct was that of the flowers and snow, but Lady Q discovered that dreams of flowers and snow were considered as symbolic of death by members of her family.'

Mr Saltmarsh comments: 'It seems to me that it is a particularly interesting dream. Not only are the details of the precognition very full and numerous, but the fact of recurrence makes it specially impressive. Also it is an example of long-distance precognition, the first dream having occurred six years before the event.'

In cases of this sort, where there is a wealth of veridical detail including an item of an unusual sort (the dead man's hand knocking against the banisters as he was carried upstairs), the alternative hypothesis of chance coincidence becomes very strained.

Another 'precognitive' dream, taken from *Proceedings of the S.P.R.*, v, 322, presents very curious incidental details.

'Mrs Schweitzer dreamed that she saw her younger son, F, with a stranger, on some cliffs. Her son suddenly slipped down the side of the cliff. She turned to the stranger and said, "May I ask who you are and what is your name?" He replied, "My name is Henry Irvin." Mrs Schweitzer then said, "Do you mean Irving the actor?" and the stranger replied, "No, not exactly, but something after that style." On waking she was very worried by the dream and told her elder son, begging him to recall his brother, who was away travelling for the firm on business. He ridiculed the matter, saying that F was quite safe as he was in Manchester.

'About eight days later F was killed on the cliffs at Scarborough, where he had gone for a week's holiday after completing his business in Manchester. Mrs Schweitzer, on visiting the place, met the man who had accompanied him on the fatal occasion and recognized him as the stranger of her dream. She inquired if his name were Henry and, being told that it was, recounted her dream. He then said that he used to recite at concerts, etc., and was always introduced on such occasions as Henry Irvin, Jr. His real name was Deverell.'

This dream well illustrates the curious and inexplicable elements so often encountered in psychical research. On what hypothesis can one explain the telepathic transmission not of the companion's real name but of a nickname connected with relatively unimportant events in his own life? Why does this irrelevant detail about a stranger take a central position in a dream foreshadowing the fatal accident to Mrs Schweitzer's son? Since it was knowledge unknown to her son (at the time of the dream the son and Mr Deverell were not acquainted), it was, presumably, somehow derived from the mind of this stranger. The curious manner in which the fact is conveyed in the dream should also be noted. It is puzzles of this sort which make this subject one of extraordinary interest to those who study it.

What attitude should we adopt towards such dreams? The acceptance of precognition implies first, that the future, or parts of the future, are foreordained; and secondly, that we are capable of foreseeing them. This subject might be dismissed as a popular myth except that an increasing number of scientists are not only interesting themselves in the matter, but there are some, curiously enough physicists rather than psychologists, who maintain that the belief in precognition is not at all inconsistent with modern theories of time and space.

It may be maintained that we must absolutely establish the facts before we begin to resort to any theory. That sounds reasonable, but this argument is not so convincing as it sounds, for if a rational and easily accepted theory can be advanced explaining how such a thing is possible, then it is far less necessary to collect evidence that it does happen. The simpler the explanation of the 'facts' the less need to verify them in detail. If a man tells us that he travelled from London to Brighton in an hour by train, we accept the fact without investigating it, because we know it is possible. Thus we have little difficulty in accepting the stories told by Mr Brighten concerning what happened on board ship since they are capable of so simple an explanation as subconscious hyperaesthesia. But if the facts do not conform to any known explanation, in other words, if they are paranormal, then it is much more important to collect as many cases as possible, submit them to the closest scrutiny, and then explain them, if we can, by as simple a hypothesis as possible.

So far, let us admit that, in spite of the large amount of material available, neither the facts nor the theory of precognition in dreams have been completely established to the satisfaction of all the experts. As regards the facts, a common error is to regard a large number of probables as constituting proof, whereas there may be some fallacy underlying all of them. A few fool-proof cases, if they can be established, are worth a mass of unproved cases however authentic they may appear. This latter occurs in 'telepathy', for many people have experience of such cases, but comparatively few are well-authenticated. On the other hand, if only a few cases are accepted as authentic, it is always possible that we have missed some vital factor. That is why in science we must have a large number of cases all well-authenticated in order to prove a point. At the same time the fact that we have not scientifically proved a point does not mean that it is not true. A 'belief' is the acceptance of a supposed truth in the absence of proof. There are some who believe in telepathy, premonition, or spiritualism, although they are not yet convinced that it is scientifically proved, and they have a right to their belief which may one day prove to be true.

First, then, as regards the facts. Hundreds of cases have been scrutinized and sifted with the greatest care by the Society for Psychical Research, which, as we have mentioned, has collected together an immense literature on dreams in relation to such paranormal material. Regarding the work of the Society, William James, the world-renowned doctor, philosopher, and psychologist, wrote: 'Were I asked to point to a scientific journal where hard-headedness and never-sleeping suspicion of sources of error might be seen in their full, I think I should have to fall back on the *Proceedings of the Society for Psychical Research.*' (*The Will to Believe*, London, 1897.)

Laboratory experiments into telepathy, which are conducted in the most stringent and watertight conditions, and which have yielded results appearing to be genuinely telepathic or precognitive, and have not been satisfactorily explained by any other hypothesis, have been described in the Soal-Goldney report entitled 'Experiments in Precognitive Telepathy' (*Proceedings of the S.P.R.*, xlvii, 1943). But with such evidence we are not for the moment concerned, since our subject is related only to dreams.

In every case we have quoted, the evidence of the dream was confirmed either by its leading to specific action (as in Mr Brighten's case) or being mentioned to others prior to the events themselves (as in my mentioning the case to the other doctor in the case of the naval rating). The dream of Lady Q was well authenticated, for it was mentioned and acted on (by extracting the promise) long before the event. This precaution is taken to exclude lapses of memory, for a person after hearing of an event may report that he had a dream of it before it occurred, whereas in fact the dream came later. Therefore it is necessary to make sure that the dream came before and not after the event, and that is best secured by the dream being reported to others before the facts were known. The theory of lapses of memory cannot therefore account for such cases as we have mentioned.

Another theory, however, in which memory is involved is that there normally pass through our minds in dreams innumerable possible happenings, the vast majority of which go unnoticed because they do not occur, but that when one of them is fulfilled we remember the previous dream relative to it and regard this as premonitory. This may cover simple dreams such as that mentioned of the naval rating, except that he had no recollection of any previous dreams of his brother's death. Dreams (with corroborative evidence) which are complex and contain considerable detail, like those of Lady Q and Mrs Schweitzer, can hardly be accounted for in this way.

I have already alluded to the alternative hypothesis of chance coincidence, and it is, of course, open to anybody to prefer this explanation. Certainly chance does some queer things; such as in the case of the lady who was passing through the City of London and was stopped by a policeman who had the same letters and number as the car she was driving. This is a striking case of pure coincidence, but to make it correspond with the cases quoted, she would have had first to dream beforehand that the occurrence took place, then have mentioned the fact to others to exclude lapse of memory, and have finally experienced the event in fact. Only the last was fulfilled.

By a very great stretch of the imagination the dreams recounted, like other well-authenticated cases, might be regarded as due to 'chance'. But there comes a point where the details involved are so accurate, and the mathematical probabilities against chance are so great, that it might be maintained that it is those who believe that these occurrences are due to chance who are the credulous ones! Even in the comparatively simple case of Mr Brighten, although he would naturally have constantly in mind all possible dangers, he was not perpetually dreaming of these dangers, still less were such dreams so vivid as to cause him to get up, and when he did get up find the dreams to be true. Chance is not impossible in such a case, but it is extremely improbable. Telepathy can account for many of these paranormal experiences if telepathy is accepted as a hypothesis, or if it can be substantiated. But neither hyperaesthesia nor telepathy can explain the dreams of Lady Q, for the events happened six years after the dream.

The idea that science is opposed to prediction as such is of course untrue. The study of astronomy enables us to predict the appearance of comets and the time of the rising sun tomorrow morning. Indeed science has been described as the power to predict results. What is denied is that the human mind is capable of predicting events which are not based on previous experience, or on such scientific evidence of the course of events, or on rational deduction: such predictions, for instance, as the prophecy that Hitler would die on a certain date or that a ship will be wrecked on the next voyage. The latter we might predict because we know that the timbers are rotten, and we know from past experience that such a ship cannot weather a storm. But that is not what is meant by a 'prediction'. 'What we want to discover is whether there are any cases where knowledge of the future, which is not based on inference from knowledge of the past and present, is ever given to man,' says H. F. Saltmarsh.

Another argument brought against premonition is that of inconceivability. Let it be said at once that the argument of the 'inconceivability' of premonition, or of anything else for that matter, is quite unacceptable. If anyone claims that he will not accept anything which is not proved and that supposed phenomena such as telepathy or precognition are inconceivable or impossible, it is up to him to prove how and why they are impossible. If he cannot prove why it is impossible, he is breaking his own principle and 'believing' something which he cannot prove. For if we ask such an objector why it is inconceivable, he has no answer; it simply means that *he* cannot conceive it, or that it does not fit in with his preconceived ideas; but that does not invalidate it. Many scientific discoveries are made by men observing things in their experiments which 'ought not to have happened', and many commonplace scientific facts of today were 'impossible' to an earlier generation. What 'ought not' to have happened may simply be something that has not previously been observed. 'It is ridiculous, impossible', said the doctors in Vienna who went straight from their operations on abscesses to deliver a woman of a child, to their colleague's contention that they carried the infection on their hands, and that simply washing your hands could prevent or reduce the awful diseases experienced in the hospitals by women in childbirth. Even when by washing his hands he drastically reduced the maternal mortality in his wards they refused to accept the facts. But the principle of asepsis is now a commonplace of surgery. Many other examples can be quoted of the unscientific bias of certain scientists of their day, and their unwillingness carefully to examine questions which did not fit into the scientific outlook of the moment and which led into untrodden paths. Let me quote two other examples, which are amusing to look back upon now, but which indicate how blind scientific leaders may be to advances in thought and achievement which are only just round the corner, on the grounds of their 'inconceivability'.

Sir William Barrett writes as follows in the S.P.R. *Proceedings*, xxxiv, 291: 'I happened to be staying in Edinburgh with that famous physicist, Professor Tait, when the news of the discovery of the telephone came to us by cable. I asked Tait what he thought of it. He replied, "It is all humbug, for such a discovery is physically impossible." '

Canon Anson, Master of the Temple, has quoted other examples in *The Truth about Spiritualism* (London, 1941).

Flammarion, the French astronomer, relates that he was present when the French Academy of Sciences investigated Edison's phonograph. One learned scientist seized the demonstrator by the collar crying, "Wretch! we are not to be made dupes of by a ventriloquist." This was in 1878. The same man, six months later, after having carefully examined the phonograph, again declared that "it was impossible to admit that mere vile metal could perform the work of human phonation." The phonograph, according to his idea of it, was nothing but an acoustic illusion – it did not exist because it could not exist; it was inconceivable, therefore it could not be. Such is the attitude of many people towards telepathy and psychical research in general.

But the question of precognition still remains a very awkward problem, for it is not simply a question of establishing the facts, but depends on the problem of the nature of Time, a problem for which neither philosophers nor scientists have yet found a solution, and certainly a problem we are not competent to deal with. This much, however, may be said. Time in the ordinary sense in which we use it is artificial; for convenience' sake we divide it into hours, days, and months. But more strictly speaking *Time is our experience of a succession of events* – and we should emphasize the word 'experience'. We say we see a star in a certain position in the sky, and so it is to us. But the position in which we 'now' see the star is one which it occupied thousands of light years ago. Moreover, it is actually now in a position which will be experienced by an earth viewer thousands of years hence. We must distinguish, then, between the event and our *experience* of the event. Anyone who could see such an event at the moment it happened would appear to the rest of us to have a 'premonition' of what others only experience later.[1]

Precognition sounds incredible because we think of time in terms of such artificial measurement of time. But if we think of time as an *experience* – 'the experience of the succession of events', there appears to be no reason why some people should not experience the same events at different times, and therefore some people experience events before other people. So 'Time, like an ever-rolling stream, bears all its sons away', but it does not necessarily bear them away at the same rate, and they may experience the same events at different times. The point is, we are not conscious of the actual *facts* of life, we are only conscious of experiencing these facts, and some people may, especially in dreams, live through experiences which have not yet 'happened' so far as they are consciously concerned, though they may have been experienced in some other, as yet unrecognized, time dimension appreciated by the subconscious mind.

A more comprehensible theory is the supposition that the subconscious mind has a wider range of vision than the conscious mind. We have already seen that the subconscious processes can solve problems which the conscious is incapable of solving on account of its greater *insight*. There seems, therefore, no reason why it should not also have greater *foresight*. The function of the conscious mind appears to be to bring to bear upon events a higher degree of concentration, but in doing so it perhaps loses in *range* of vision – for the more we concentrate the less aware are we of what is going on in the periphery around. The subconscious mind, while less distinct and concentrated, may retain the wider vision, and can therefore see the solution of problems, especially by similarities and analogy, which are beyond the range of the

1. 'We must distinguish between two sorts of time, or rather between two different conceptions of it, viz., the mathematician's and physicist's time, and psychological time. The former is admittedly an abstraction; it consists of point-instants having no magnitude but only temporal position; it has no more real existence than the mathematical point or line. Psychological time is time as experienced. What is really the object of experience is change, *i.e.* events taking place. Time is an abstraction from change.' (Saltmarsh, *Proceedings of the S.P.R.*, xlii, 1934.)

conscious – just as there are some lower animals whose eyes are so set that they have a much wider range of vision than we have and can see events which are happening behind them, of which we are quite unaware, as well as in front. Thus our total range of awareness may not be strictly limited to what is in the full focus of consciousness, but may spread to a much wider field than that which is in the more concentrated attention of the conscious mind. The captain on the bridge of the *Queen Elizabeth* can see land which is quite hidden to the man on the lower deck, who might deny, therefore, that there is any land in sight because he cannot see it. Thus some people may have a premonition of events denied to others.

Saltmarsh has advanced such a theory, and suggests that what we call 'the Present' is not a point without magnitude but carries duration – the 'specious present'. If, he says, we imagine ourselves looking through a pinhole at a moving cinema picture, we see only a limited field: a person looking through a bigger hole sees a wider range of the picture and therefore *knows what is coming* on the screen before the person looking through the smaller hole, and may tell him so. If the man looking through the smaller hole did not know the explanation, he would suspect the other of having a premonition. So the conscious mind, whose function it is to deal with the practical present, may have a much more limited 'present' than the subconscious, which may have a more extensive range of experience. In such a case the subconscious mind sees what is coming before the conscious mind: but as there is intercommunication between the conscious and the subconscious minds (as in dreams), the subconscious mind may be able to make the conscious mind aware of what is coming, and this we call a premonition.

It may be that the subconscious mind is *always* aware of this wider range of knowledge, but because it is not in the full focus of consciousness we are not usually aware of it and do not usually make deliberate use of it, because the conscious mind, especially in our waking state, is so concentrated on its present. But when the conscious mind is in a state of relaxation (as in the cases mentioned on page 113 regarding scientific discoveries) or asleep, the subconscious mind may be able to convey to it the things it has sensed. That is what constitutes our intuition. The intuitive person is the one who makes most use of subconscious perception and subconscious inference. That is why the scientist sees relationships and discovers things when he relaxes which he cannot see when he concentrates: it may also explain how people of the logical 'thinking type' as described by Jung are often deficient in intuition. There are many things that are hidden from the wise and prudent which are revealed to babes.

Coming back to our main problem, if we accept that it is possible that the subconscious processes of the mind, which are active during sleep, have a wider range of perception than the logical conscious mind, we could explain phenomena like the captain's dreams of the collision and of the parting of his mooring ropes by this theory, for in his sleep and by a subconscious hypersensitivity he was able to foresee what was going to happen, although in his waking state he was unaware until he went aloft of what was coming. Similarly the naval rating in his sleep may have had a far wider range of perception than in his waking life, and therefore knew what was happening to his brother in France. Not only so, but if we accept Saltmarsh's analogy (and he does not claim that it is anything but an analogy), it would suggest the possibility of precognitive intuition in waking life, such as that of impending disaster which prevents someone travelling by a certain ship or train, as well as the precognitive dreams mentioned. We must not therefore categorically reject the possibility of precognition until we know more of what we mean by Time, and more of the nature and functions of the subconscious mind. These are only analogies and prove nothing, but such possibilities may help to make us less dogmatic in our assertions that certain things like precognition cannot happen.

My own attitude regarding these psychic phenomena, for what it is worth, can be stated quite simply. If I am asked whether I am convinced of such phenomena, my reply is that as I am incapable of seeing through the tricks of the modern conjurer, which were acknowledged to be simply tricks, I do not feel myself competent to pass judgement as to whether these phenomena of precognition, telepathy, mediumship, and spiritualism were genuine or not; they might easily be tricks and I not know it. These are matters for the expert in psychical research, and this includes a knowledge of mathematics, physics, and psychology, not to speak of an intimate acquaintance with the principles of conjuring and the practices of the charlatan.

We may now hark back and ask how do the dreams we have mentioned in this chapter relate to the biological theory we have propounded. Quite simply! There is not one of these dreams which is not associated with an acute problem in the dreamer's mind. The dreams of Mr Brighten in his yacht obviously related to very serious problems of the safety of his ship in the fog and of parting from his mooring. The dream of the naval rating similarly related to a worry about his brother in France, so that his mind was already attuned for the reception of the news when it came. The other precognitive dreams are equally concerned with worries and problems in the mind of the dreamer, all of which press for solution. In all of them, however, no solution is offered, unless we regard Mr Brighten's waking up and going on deck as such. They only allude to disturbing facts.

We therefore conclude as we began: that the function of dreams is, by the reproduction of the problems of the day, subjective or objective, to work towards their solution. In some cases they go no further than calling attention to the problems: in other cases they lead to suggestions of a solution which are tentative and inadequate. In other cases they offer real solutions to both the moral and practical problems of life. For that reason they demand our serious study and attention and they will, no doubt, continue to have a fascination for man such as they have had all down the ages.

LIST OF DREAMS QUOTED

INDEX OF AUTHORS

GENERAL INDEX